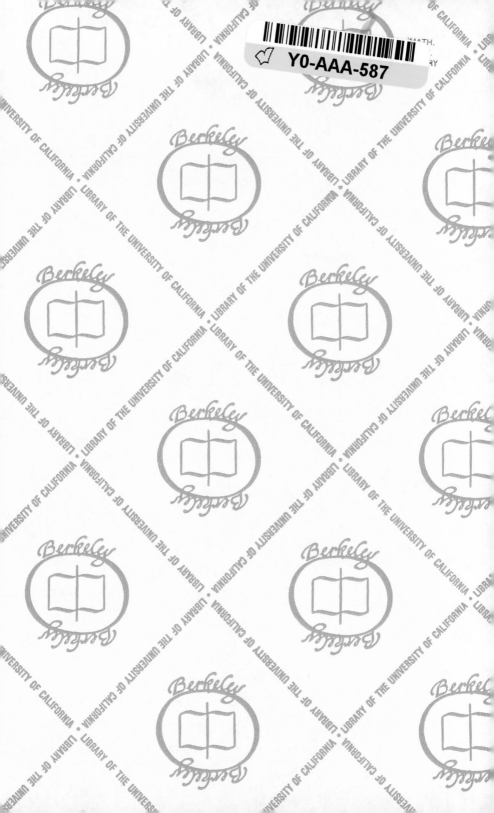

Y0-AAA-587

MATH.
STAT.
LIBRARY

Glenn Satty, Thomas J. Blakeley, James G. Colbert
Computing and Logic
Mathematics and Language

Introductiones

Contributions to Philosophical Analysis

Editor
Hans Burkhardt · Erlangen

Philosophia Verlag München Wien

Glenn Satty · Thomas J. Blakeley · James G. Colbert

Computing and Logic Mathematics and Language

Philosophia Verlag München Wien

o 3753025
MATH-STAT.

CIP-Titelaufnahme der Deutschen Bibliothek

Satty, Glenn:
Computing and Logic: Mathematics and Language / Glenn Satty; Thomas J. Blakeley;
James G. Colbert – München Wien : Philosophia Verl., 1988.
(Introductiones)
ISBN 3-88405-071-0

MYLAR
MATH

ISBN 3-88405-071-0
© 1988 by Philosophia Verlag GmbH., München
All rights reserved. No part of this book may be reproduced in
any manner, by print, photoprint, microfilm, or any other means
without written permission except in the case of short quotations
in the context of reviews.
Manufactured by Druckhaus Beltz
Printed in Germany 1988

CONTENTS

QA9
S317
1988
M ATH

Preface

We would like to thank all who have collaborated in bringing this volume to press, including Timothy W. Blakeley who did the artwork. Bob and Christine Gilman supplied logistic support for Glenn during his frequent visits to the Boston area.

The present work is obviously a textbook for persons interested in computing and logic. Lest the combination appear too formidable, the book could be used to study logic alone or even learn BASIC alone. There are probably large groups who feel comfortable with logic or computing already, but suspicious of the other. We hope that the book encourages them to take the plunge.

Some Introductory BASIC

0.1   Constants and Variables

A *constant* is an expression whose value can never change, i.e. it remains constant, while a variable is an expression which can take on any value from among given permissible values, i.e. it can vary within a given domain. In mathematics, expressions such as 12, -2.781, 3/5, pi, and sqr(2) are constants, for they all have only one possible value, while X, Y, P, A, I represent variables. In the BASIC computer language, we distinguish between 3 types of constants and, correspondingly, 3 types of variables.

(1) *Integers*. An integer is a whole number, either positive, negative, or zero. We always use a % sign to represent an integer. 5%, -8%, 11532%, 0% are all integer constants. So we may speak of 5% marbles, of -8% trees missing from a plot, of 11532% people attending a concert, since marbles, trees, and people are integer concepts (one never speaks of 1/2 a person at a concert, though one may not be 'all there'!).

A letter followed by a % sign, such as A%, X%, p%, represents a variable which can take on only integer values. An integer variable can be longer than a single letter, such as P1%, FRED%, J7235%, but must always begin with a letter and end with a % sign. Note that the % sign has nothing to do with "percentage"--it merely indicates the presence of an integer: variable or constant.

(2) *Reals*. Most numbers familiar to us all are real constants, such as 7, 3.25, -.0072, 1/2, pi, and sqr(2), which are simply real numbers. We see that all integers, such as 7, are real numbers; but not all real numbers are integers. A real variable is an expression which begins with a letter: P, R2D2, FRED. All of these variables represent real numbers. Integers and reals are referred to as *numeric* variables or constants.

(3) *Strings*. The most important kind of constants and variables which we will be using are *strings* or *alpha-numerics*. A string constant is any expression which is enclosed within quotes: "HELLO", "A tree grows in Brooklyn", "P", "FRED",

1

"3.25". The value of these constants is the expression itself. It may seem strange that words and sentences are considered constants, but they are expressions whose values never change! Note that any real number may be a string constant, as long as it is contained within quotes--in which case it no longer has a "numeric" value.

The corresponding string variables begin with a letter and end with a $ (which has nothing to do with money), such as: P$, J25$, FRED$--all of which must take on string values. One must be careful with these variables and constants, because a slight change in an expression may cause a radical change in its meaning, for example:

P%	integer variable
P	real variable
P$	string variable
"P"	
"\"	
"P%"	are all string constants

One Final Note: Different machines have different restrictions on the length of the variable names and permissible symbols in variable names. See your user's guide for the specifics.

0.2   Assignment Statements

In all computer programs we will want to assign values to variables. The simplest assignment is the LET statement. For example, the statement

100   LET P = 2.5

assigns to the real variable P the value 2.5 which it retains until it is assigned a different value. Note that 100 is a line number. In many implementations of BASIC, line numbers are required.

Variables must be assigned their proper type of value. That is, string variables must be assigned string constants, real variables must be assigned real constants, and integer variables must be assigned integer constants.

2

The following statements are *legal*:

```
500      LET C% = 17%
600      LET R2D2 = 3.1415
235      LET JANE$ = "HIHO"
```

From statement 235 we know that whenever we use the variable JANE$, it will have the value "HIHO". If, later in the program, we say

```
980      LET JANE$ = "TATA"
```

its value changes to "TATA".

The following statements are *not legal*:

```
200      LET C%=15.7
1000     LET R2D2 = "HIHO"
1530     LET JANE$ = HIHO
```

Statement 200 will not cause an error message, but it will cause C% to have the value of 15% since it must be an integer (and to make C% an integer the computer will "truncate" or chop off the decimal part), so that programming calculations will be off. Statements 1000 and 1530 will cause the program to stop processing, and the error "ILLEGAL MODE MIXING" will be printed out, since 1000 tries to assign a string constant to a real variable, and statement 1530 tries to assign a real variable (HIHO) to a string variable.

Other legal statements include:

```
650      LET D = B+A
750      LET C2$ = JANE$
800      LET C% = C% + 1%
```

In 650, D is assigned the value of the sum of the values of B and A. In 750, C2$ is assigned whatever value JANE$ represents (if it is still "HIHO" then C2$ will also be "HIHO"). 800 is the strangest of all! C% is assigned a value equal to its *old* value plus 1. This type of statement is called a counter since it is used to count the number of times a process has taken place. We will discuss counters later in this chapter.

The general rule of the assignment LET is:

The expression on the *left* of the equal sign must be a variable and is assigned the value which is represented by any valid combination of variables "of the same type" on the *right* of the equal sign.

The LET statement is an internal assignment statement; i.e. it assigns values from within the program. BASIC also has external assignment statements, the most common of which is INPUT. The INPUT statement permits us to assign values from outside the program to variables which are written within the program. The statement:

```
300      INPUT A
```

will cause a ? to appear at the computer terminal. The ? is prompting the user to enter a value for the variable A, to be used within the program. Once the value is typed in and the carriage is returned, A is *assigned* the entered value. Thus INPUT is a LET statement in which the user can choose the value of the variable.

Question: What is wrong with this series of statements?

```
200      INPUT P$
201      LET P$="RIGHT ON"
```

Answer: The INPUT statement is meaningless, since the value of P$ is immediately reassigned the value "RIGHT ON".

Output: END, RUN, REM

In order for a user to gain access to the results of any program, the results must be output in some fashion. This is most commonly achieved with a PRINT statement, which can print both variables and constants.

```
500      LET A% = 5%
510      PRINT "THE VALUE OF A IS ";A%
```

Line 510 will cause the computer to print the constant (which is contained in the quotes) followed by the value 5 (which is the value of the variable A%) when the program is RUN. Any

program can be RUN once it has been terminated with an END
statement, after which *no further program statements may follow*.
The standard sequence is:

```
500      LET A% = 5%
510      PRINT "THE VALUE OF A IS ";A%
1000     END
RUN
```

THE VALUE OF A IS 5

Notice RUN does not have a line number because it is a *system
command* rather than a BASIC programming statement, as are
LET, PRINT, INPUT, and END. In addition, we see a
"separator", i.e. semicolon, between the two items printed out.
There must always be a semicolon or comma between different
variables or constants in a PRINT statement. A semicolon
produces printouts which are next to each other, while a comma
produces printouts which are in separate registers (columns). In
our statement 510 we have a space inside of the quotes so that
there will be a space between *is* and 5. If we were to change the
semicolon to a comma, the printout would be something like:

THE VALUE OF A IS   5

The following program is designed to INPUT a student's
name, two testscores, and PRINT the student's average.

```
300      INPUT N$,A,B
350      PRINT N$;"'S AVERAGE IS  "; (A+B)/2
500      END
```

When we RUN the program, a ? appears. We type "LUCY", 85,
92. The computer then prints:

LUCY'S AVERAGE IS 88.5

We note the following about the program and the printout:
1) We can print out a string variable, a string constant, and
    the result of a simple computation with one PRINT state-
    ment.
2) The printout appears on a single line, nicely formatted.

5

3) A+B must be in parentheses since A+B/2 is interpreted by the computer as A + (B/2) = 131. This is so since the order of arithmetic operations is:
within parentheses, from left to right

   i)   EXPONENTIATION($\wedge$ or **)
   ii)   MULTIPLICATION(*) or DIVISION(/)
   iii)   ADDITION(+) OR SUBTRACTION(-)

4) The ? does not indicate what is to be INPUT. This can be remedied by a simple PRINT statement before the INPUT, such as:

290   PRINT "TYPE IN A NAME AND 2 TESTSCORES PLEASE";
                 (we end with a ; so that the ? appears on the same line)

or by changing the INPUT to read

300   INPUT "TYPE IN A NAME AND 2 TESTSCORES
               PLEASE";N$,A,B

which is permissible in most versions of BASIC.
5) Nothing in the program tells us what the program will be about, or who wrote it, when, etc. This can be remedied by a REM statement or a !, both of which indicate that a remark and not a program command follows.
6) We could have calculated the average in a separate statement, such as

310   LET AVE = (A+B)/2

and then printed AVE in the final PRINT statement.

Considering the above, a nice program would be:

90   ! PROGRAM AVERAGE
100   ! GLENN SATTY   JULY 4, 1987
110   ! THE PURPOSE OF THIS PROGRAM IS TO CALCULATE
120   ! THE AVERAGE OF A STUDENT WHOSE NAME AND
130   ! TESTSCORES ARE INPUTTED AT THE TERMINAL
300   INPUT "TYPE IN A NAME AND 2 SCORES PLEASE";N$,A,B
310   LET AVE = (A + B)/2

```
320     PRINT N$;"'S AVERAGE IS ";AVE
500     END
```

0.3   Branching

The above program is fine for calculating one student's average, but is terribly inefficient for a large number of students, since we have to continually type RUN for each student. A better way to accomplish our task is with an unconditional branch or GOTO statement at line 330:

```
90      ! PROGRAM AVERAGE
100     ! GLENN SATTY    JULY 4, 1987
110     ! THE PURPOSE OF THIS PROGRAM IS TO CALCULATE
120     ! THE AVERAGE OF A STUDENT WHOSE NAME AND
130     ! TESTSCORES ARE INPUTTED AT THE TERMINAL
300     INPUT "TYPE IN A NAME AND 2 SCORES PLEASE";N$,A,B
310     LET AVE = (A + B)/2
320     PRINT N$;"'S AVERAGE IS ";AVE
330     GOTO 300
500     END
```

The statement at 330 causes an unconditional branch to line 300: go directly to 300 (do not pass GO, do not collect $200).

All programs run in numerical order unless a branching statement, as in 330, occurs. When this program is RUN, the following sequence of events transpires:

```
TYPE IN A NAME AND TWO SCORES PLEASE?
    "LUCY",85,92
        LUCY'S AVERAGE IS 88.5
TYPE IN A NAME AND TWO SCORES PLEASE?
    "CHANG",88,98
        CHANG'S AVERAGE IS 93
TYPE IN A NAME AND TWO SCORES PLEASE?
    "BOY GEORGE",74,82
        BOY GEORGE'S AVERAGE IS 78
                    .
                    .
            . (forever)
```

The problem now is to get the computer to stop! The program has caused an *infinite loop* which can only be interrupted with a system control; e.g. by hitting the ctrl button and C button simultaneously--^C, which is quite awkward. (Some systems have "disaster" controls other than ^C. Check the user's manual.)

A better solution to this problem is with a "conditional branch", which has the form:

305   IF (*expression*) THEN (*execution*)

where (*expression*) stands for any statement which is either true or false and (*execution*) stands for any executable statement, such as GOTO, INPUT, PRINT, LET, or another IF statement. The executable statement is executed when the expression is *true*.

In our program we will set up a *flag variable* which indicates the end of our input. This will be the word "DUMMY". We will also add a PRINT statement to format our output in a better way, and introduce a counter to count the number of students which have been averaged. Our final program looks like this:

```
90      ! PROGRAM AVERAGE
100     ! GLENN SATTY    JULY 4, 1987
110     ! THE PURPOSE OF THIS PROGRAM IS TO CALCULATE
120     ! THE AVERAGE OF A STUDENT WHOSE NAME AND
130     ! TESTSCORES ARE INPUTTED AT THE TERMINAL
200     LET C = 0                    ! INITIALIZE THE COUNTER
300     INPUT "TYPE IN A NAME AND 2 SCORES PLEASE";N$,A,B
305     IF N$ = "DUMMY" THEN GOTO 400      ! EXIT THE LOOP
310     LET AVE = (A + B)/2
320     PRINT N$;"'S AVERAGE IS ";AVE
325     LET C = C+1   ! INCREMENT COUNTER BY ONE STUDENT
327     PRINT       !PRINTS A BLANK LINE BETWEEN STUDENTS
330     GOTO 300
400     PRINT C; " STUDENTS HAVE BEEN AVERAGED"
500     END
```

and the RUN might look like:

```
TYPE IN A NAME AND TWO SCORES PLEASE?
    "LUCY",85,92
        LUCY'S AVERAGE IS 88.5
```

8

TYPE IN A NAME AND TWO SCORES PLEASE?
   "CHANG",88,98
      CHANG'S AVERAGE IS 93

TYPE IN A NAME AND TWO SCORES PLEASE?
   "BOY GEORGE",74,82
      BOY GEORGE'S AVERAGE IS 78

TYPE IN A NAME AND TWO SCORES PLEASE?
   "DUMMY",0,0

      3 STUDENTS HAVE BEEN AVERAGED
Ready

and the program comes to an end.

Let us note that the most elementary true or false expressions have one of the following relation signs:

SIGN	MEANING
=	EQUAL
<	LESS THAN
<=	LESS THAN OR EQUAL
>	GREATER THAN
>=	GT OR EQUAL
<>	NOT EQUAL

Examples:

$5 < 7$ is TRUE
$16 >= A$ is ? unless we know the value of A
G$ <> "DUMMY" is ? unless we know G$
"ALABAMA" > "ALASKA" is FALSE since strings are ordered alphabetically, with A's first and Z's last

0.4   Formal Looping

The loop created in 300 to 330 of the last program is a bit contrived. Although it accomplishes our task, one prefers not to use GOTO's unless absolutely necessary. Many early programmers and "hackers" tended to GOTO all over the place, making programs both inefficient and unreadable. One method of avoiding GOTO is a FOR...NEXT loop. Let us say that we know that there are exactly 10 students in the class. We create the following program:

```
90        ! PROGRAM AVERAGE
100       ! GLENN SATTY   JULY 13,1987
110       ! THE PURPOSE OF THIS PROGRAM IS TO CALCULATE
120       ! THE AVERAGES OF 10 STUDENTS INPUTTED AT
130       ! THE KEYBOARD
290           FOR I% = 1% TO 10%
300           INPUT "TYPE IN A NAME AND TWO TESTSCORES
                   PLEASE";N$,A,B
310           LET AVE = (A + B)/2
320           PRINT N$"'S AVERAGE IS ";AVE
325           PRINT !PRINTS BLANK LINE BETWEEN STUDENTS
330           NEXT I%
500       END
```

The loop lies between 290 and 330 and is indented for readability (indenting does not matter to the computer). The loop is proceeded through 10 times, as I% takes on values from 1% to 10%. That is, the first time through, I%=1%, the next time I%=2%, the next I% is 3% ... until I%=10%. When I% = 10%, the next time we get to NEXT I% the loop is exited and the first statement which follows the loop is executed--in this case the program ENDs. Notice that the loop has its own built-in counter (I), which can be used within the loop but whose value should not be reassigned within the loop. The standard form of a FOR...NEXT... loop is:

```
1000      FOR I = A TO B STEP C
              .

              .
1500      NEXT I
```

Example:

```
1000      FOR I = 2 TO 12 STEP 3
1100          PRINT 2*I,
1500      NEXT I
```

10

This outputs:

       4   10   16   22

The columnar spacing in the output is due to the comma at the end of the print statement. I takes on the values of 2, 5, 8, and 11, since it increases by 3 every NEXT (that's what the STEP is all about). Note that STEP may be any real number, including fractions and negatives, as in FOR I = 20 TO 18 STEP -.5, in which I would take on the values of 20, 19.5, 19, 18.5, 18 successively each time it progresses through the loop.

0.5   Subscripted variables I: One-Dimensional Arrays

Often, especially when dealing with loops, one needs to work with many variables which have similar but "varying" names. The task is accomplished with *subscripted variables*. For example, if we want to keep track of our ten students' names, we can use the following 10 assignments:

N$(1) = "LUCY"   N$(6) = "BENITO"
N$(2) = "CHANG"   N$(7) = "MAUREEN"
N$(3) = "BOY GEORGE"   N$(8) = "HEPZIBAR"
N$(4) = "ROSANNA"   N$(9) = "MARIA"
N$(5) = "JASCHA"   N$(10) = "SEAN"

The following loop:

```
500     FOR I% = 1% TO 10% STEP 2%
510         PRINT N$(I%)
520     NEXT I%
```

would print:

LUCY   BOY GEORGE   JASCHA   MAUREEN   MARIA

which are the values of N$(1), N$(3), N$(5), N$(7) and N$(9).
    The *subscript* or index (the number inside the parentheses)

must always be an integer, though it need not have a % sign. The variable which is subscripted, in this case N$, can be a string, a real number, or an integer. Thus P%(5) would have an integer value, P(5) would have a real value, and P$(5) would have a string value.

In order to use a subscripted variable, one must reserve computer space for such variable. This is done with a DIM statement, as in

100   DIM N$(10)

Statement 100 creates a variable N$ with 10 different parts, each of which can be filled with different constant values. All need not be filled, but not more than 10 may be filled. If one tries to assign or access N$(11) a "Subscript out of range" error occurs.

N$(1) N$(2) N$(3) N$(4) N$(5) N$(6) N$(7) N$(8) N$(9) N$(10)

Before assignment, all of the above variables have the value " " (or 0 (zero) in the case of reals or integers). After the above assignment, we have:

LUCY	CHANG	BOY GEO.	ROS-ANNA	JAS-CHA	BEN-ITO	MAU-REEN	HEP-ZIBAH	MARIA	SEAN

N$(1) N$(2) N$(3) N$(4) N$(5) N$(6) N$(7) N$(8) N$(9) N$(10)

We will be using subscripted variables throughout our text. When the time comes, we will introduce two-dimensional arrays, which have two indices, as in N$(I,J). Our single indexed arrays are called *one-dimensional arrays*.

0.6   Subroutines

Many of the programs that we will write in this book employ an identical series of operations. In order to alleviate the need constantly to rewrite the same set of instructions, we use the *subroutine*, a useful device in all programming. The BASIC command for employing a subroutine is:

600   GOSUB 5000

where 5000 is the line number at which the subroutine begins. The subroutine must contain a final RETURN statement, which returns processing to the line immediately following the call (GOSUB statement). For example:

1500   GOSUB 5000
1510   (any statement)
            .
            .
            .
5000   ! SUBROUTINE BEGINS
            .
            .
            .
7000   RETURN   ! END OF SUB - BACK TO 1510

0.7   String Functions

Most of our work in this book will be with string variables. We will often want to work with elements within the string itself, as individual words within a sentence. In order to do such manipulations, we need to employ special string functions, which are standard in all BASIC dialects, though they may differ slightly in form. These include, but are not limited to:

1) LEN(A$)--is the number of characters or length of A$, including punctuation and spaces.
2) LEFT$(A$,I)--is a string of the first I (a number) characters of A$ from the left.
3) RIGHT$(A$,I)--is a string of all the letters to the right of the Ith character, which includes the Ith character.
4) MID$(A$,I,J)--is a string of length J starting with the Ith character of A$.
5) INSTR(I,A$,B$)--is the place where B$ first occurs in A$, starting to look from the Ith place.

We can also add or concatenate strings (concatenate = to link

13

together as in the linking of a chain), as in A$+B$--in which case B$ will be appended immediately at the end of A$.

The following example demonstrates the use of 2-4 and concatenation:

```
LET A$ = "MY DOG HAS FLEAS."
LET C$ = "CAT"
```

If we want to change DOG to CAT and printout MY CAT HAS FLEAS, we may do the following:

```
LET P$ = LEFT$(A$,3)          which is "MY "
LET Q$ = RIGHT$(A$,7)         which is " HAS FLEAS."
LET B$ = P$ + C$ + Q$         which is "MY CAT HAS FLEAS."
```

In the above example, LEN(A$) = 17.

We can make this example a bit more sophisticated with the use of INSTR, which assumes that we do not know the position of "DOG" in the sentence, as in the following program:

```
100     LET A$ = "MY DOG HAS FLEAS"
110     LET C$ = "CAT"
120     LET K = INSTR(1,A$,"DOG")
130     IF K = 0 THEN PRINT "DOG NOT FOUND" \GO TO 500
140     LET B$ = LEFT$(A$,K-1) + C$+RIGHT(A$,K+3)
150     PRINT B$
500     END
```

We note that INSTR = 0 means that the given string ("DOG") never occurred. Also, the \ indicates a continuation of the THEN part of the IF... THEN... statement. Note that none of the above string functions can be "assigned" values with a LET statement. That is, none of these functions can ever occur on the left side of an = sign. For example,

```
300     LET LEFT$(A$,4) = "BOOM"
```

is illegal. It also makes no sense, since LEFT$(A$,4) is already *defined* as the first four characters of A$.

When using string functions, one must be very careful with spaces, and with beginning and ending letters.

0.8   Algorithms

Algorithms are employed unconsciously throughout our daily lives, and consciously in our computer programming. An algorithm is a set of directions which gets us unambiguously from an initial state to a final state.

Examples of algorithms abound. They include such varied ideas as directions for getting to the store from your house, directions for turning on a burglar alarm, instructions for finding the square root of a number, and directions to the computer of a missile for the target to be hit.

To reiterate, there are three things which we must understand about algorithms:

1) The initial state, or starting state must be defined.
2) The final state, or end state, must be known.
3) The set of directions must be unambiguous. That is, there may not be a directive, for which a wrong choice is permitted. (A wrong choice here means one which pro-hibits reaching the end).

The definition of algorithm contains a lot of leeway. For example, although the directives must be unambiguous, it is nowhere stated that the directives must be simple or must prescribe the most direct route. One may take the "scenic" route and still achieve the desired end. Instead of crossing the street to get to the other side, one may want to go the other way, swim across the ocean, cross the continents, another ocean, another continent, and end up on the other side of the street! (Some computer programs give the impression of doing just this).

Also, we have not stated that the number of steps taken to reach the end must be finite. But, with respect to computer applications, as well as many other applications of algorithms, the only sensible algorithms are those for which the end can be achieved in a finite number of steps.

Some people also add the requirement that the algorithm take a "reasonable" time to complete--but the interpretation of "reasonable" is continuously changing relative to computer speed and storage.

Furthermore, we often look for the "best" possible algorithm --which may mean the least number of steps, the shortest time, the smallest use of storage capacity, greatest intelligibility to the widest range of people, or some combination of the above.

To illustrate the concept of an algorithm, let us formulate one for our morning socking and shoeing. The initial state is bare feet and the final state is shod and ready to walk.

Shodding Algorithm
 1. [Initial State] Barefoot
 2. Place sock on left foot
 3. Place sock on right foot
 4. Place left shoe on left foot
 5. Tie left shoe
 6. Place right shoe on right foot
 7. Tie right shoe
 8. [Final State] Shod

We can make the following observations about this algorithm:

First, the algorithm is not unique. We could easily interchange steps 2 and 3; or do steps 2, 4, 5, followed by 3, 6, 7. Of course, we cannot do step 5 before 4, or 4 before 2.

Next, we may insert the step "brush teeth" anywhere in the algorithm, although it does not make much sense in the context of a shodding algorithm.

Finally, it may seem that this algorithm is unambiguous; but, what if someone does not know how to tie shoelaces! Steps 5 and 7 are now ambiguous; and we may need an algorithm for tying shoelaces (in a subroutine). Although shoelaces are tied twice in our algorithm, we need to write the shoelace-tying algorithm only once. Then, the directive at 5 and at 7 would say "Refer to shoelace-tying algorithm", as a "procedure", a "subroutine", or a "sub-program". (All can be eliminated if you wear loafers.)

Thus, we see that in order to follow an algorithm, one must understand each directive of that algorithm, and the condition of "unambiguousness" is dependent upon the knowledge of the one carrying the algorithm out.

From what has been said in the previous sections of this chapter, we can write a "pseudo-program" (since most computers cannot put their own shoes on!), using the ideas of BASIC.

Shodding Pseudo-Program

```
100      FOR I = 1 TO 2              ! a foot at a time!
110          PUT ON SOCK(I)
120          PUT ON SHOE(I)
130          GOSUB 1000             ! shoelace-tying routine
140      NEXT I
150      STOP ! prevents false entry to subroutine
1000     ! SHOELACE-TYING SUBROUTINE
             .
             .
             .
2000         RETURN                 !  returns to 140
9999     END
```

Exercise: Write an algorithm for tying shoelaces.

I
Bases

Logic is the art of manipulating the correct forms of thought. There are questions that can be asked about logic on a theoretical level--these belong to the "philosophy of logic". Logic itself is not just a science (a matter of knowledge only) but also an art--i.e. it is a way of doing something. What is done by logic is the manipulation of correct forms. What these "forms" are will come out as we go along, as they will be the main subject of our account. That they are "forms of thought" has been disputed by many philosophic schools. But, again, this is a matter that can be left to the philosophy of logic.

As an example of "manipulation" let us take the clause "the horizon is hazy". We could choose to let W$ represent this clause. Or, we could let S$ represent "the horizon", C$ represent "is", and P$ stand for "hazy". Which representation we do choose depends on what we want to do with the clause: manipulate it as it is with respect to other clauses, or analyze it in terms of its parts and their relationships with each other.

Here is an example of one of the possible manipulations:

```
10      S$ = "the horizon"
20      C$ = " is "
30      P$ = "hazy"
40      Print S$,C$,P$
50      Print C$+S$+P$
60      Print P$+C$+S$
70      Print P$+S$+C$
RUN

        the horizon is hazy
        is the horizon hazy
        hazy is the horizon
        hazy the horizon is
```

In the above analysis of the clause with respect to its parts, we find that the first and third manipulations make perfectly good sense in both grammatical structure (syntax) and meaning (semantics). The second manipulation only makes sense when interpreted syntactically as a question. The fourth manipulation can only make sense if the words have a different meaning from that which they had at the outset; i.e. "is" must be taken to mean "exists", and hazy is that which *exists*. Note that we can conceive either "hazy, the horizon", where *hazy* is the name for the horizon, or "hazy the horizon", one name, possibly that of a racehorse. Basic manipulations are the subjects of Chapter II, while the syntactic and semantic problems are investigated in Chapter III.

The "forms" that have stood out most clearly in the history of logic are the various "sentential functions"--the and(s), if(s), not(s), or(s) and is(es). It may seem strange to see these words in the plural. Is there more than one "and", "if", "not", "or" and "is"? As a matter of fact, there are several meanings--or, at least, uses--for these words. When we say, for example, "The book is red" we are attributing a property or quality (red) to a thing (book); while, when we say "A camel is a horse put together by a committee" we are speaking figuratively, and "Man is a rational animal" is intended as an essential definition. Each occurrence of "is", then, takes on different characteristics, in function of the context in which it is used--what one might call "geographical" specification of the term.

Something similar can be indicated in the case of "and". "John and Mary" are a couple of people; "ham and eggs" is a breakfast; "Punch and Judy" form a famous comedy team. Do the differences among the associated terms depend on or occasion differences between these uses of "and"? This is a typical question of logic.

One could take the approach that these logical connectives have different properties or functions, depending on whether the extremes are "things", "properties", "times", "places", etc.; and logic will want to know not only if this is the case but exactly how this occurs, if it is indeed the case; and why it is not, if it proves to be a fruitless question. In our final section, we will be constructing parts of a *Language Analyzer*, where the differences among these connectives will allow us to differentiate among the words of a sentence.

It is clear that we can say of the connective "and" that it either

connects substances (things)	or does not
connects properties (qualities)	or does not
connects quantities	or does not
connects locations	or does not
connects times	or does not
connects actions	or does not
and so on	

while there are problems involved in connecting, for example, substances and properties, or properties and quantities. One says "dog and cat", "dog or cat" quite easily. Although one can say "dog and large", "dog or large" is not as easily said.

It is also clear that "is" either

correlates places	or does not
correlates persons	or does not
and so on	

Here, too, we see that "is" works well in "dog is large" but not in "dog is cat". Similar questions can be asked all along the line.

The various "or(s)" and "if...then(s)..." are the occasion for other questions, having to do with how alternation (...or...) and consequence (if...then...) come to be for the various categorial levels we have been mentioning above. These are representable in the "branching" functions that are part and parcel of computer operation. So, for example, "He will go to work or he will not go to work" represents a use of "or" that differs from that in "He is going to smoke or to chew gum", and they raise further questions on the logical character of such terms. Also, these examples will help us see that the mutually exclusive terms can be represented in the binary code of the computer.

Throughout this course of study we will be concerned with binary-valued representations; i.e., representations to which one can attribute one and only one of two values. For example, a statement is either *true* or *false*; a process of reasoning is either *valid* or *invalid* ; an item either *exists* or *does not exist*; the light is either *on* or *off*; the coil is either *magnetized* or *not magnetized*. Although the use of binary-valued representations appears to be restrictive, we will later see that the only limitations which we encounter--whether they be human, systemic, or mechanical--are independent of our systems of binary-valued representation. That is to say that this mode of representation is no more or less restrictive than any other finitely valued representation--and it is the clearest mode of representation to understand, as well as being "machine-friendly".

Before going into the "logic of terms" in Chapter II, which is the most difficult part of the whole field of logical questions, we want to begin where most logicians have traditionally begun, and that is with the "logic of propositions" or "logic of sentences"; for, it is precisely within the context of the proposition that one can best examine the logical properties of the connectives.

The propositions in which these logical connectives occur can be classified according to their "quantity" and "quality".

"This man walks upright", "Some men walk upright", "All men walk upright"--these propositions differ as to "quantity", the first being singular, the second particular, the third universal.

According to quality, a proposition may be affirmative, "This man is eligible to vote", or negative, "This man is not eligible to vote".

This gives us six types of proposition according to their quantity and quality, four of which formed the "square of opposition" of traditional logic:

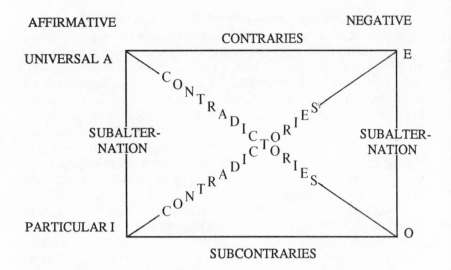

A	=	All men walk upright
E	=	No man walks upright
I	=	Some man walks upright
O	=	Some man does not walk upright

All the propositions we have discussed thus far have been "assertoric"; i.e. they assert something about something (a property about a thing, a place about a person, etc.). "Modal logic" is the name given to the set of questions that arise from propositions that say that something is "possibly", "necessarily", or "impossibly" the case (see Section 2.311).

*　　*　　*

These questions do not exhaust the domain of logic--we have not even mentioned all the various logics (there is, for example, a "dialectical logic", according to some); but they give an indication of the scope of logic. As to its tools, like mathematics, logic makes extensive use of symbols; so much so that contemporary logic is often called "mathematical logic", "symbolic logic", or "algebraic logic". Whatever else it is, logic is a "formal" discipline. It uses its symbols, as far as possible, without reference to things that are named (how, where, when, etc. they exist; whether or not they exist, and so on)--i.e. these symbols are used "formally" and the general term "formal logic" is applied to all of what we have been discussing above.

"Formalism" implies universal substitutability, iterativity and referential detachment, meaning that

(1) anything can be substituted for a symbol as long as it is put for every occurrence of that symbol;
(2) any valid formula can occur in any position in another formula; and
(3) what objects are symbolized is of no importance, as long as the substitution is consistent.

This can best be illustrated by an example from Aristotelian logic: the true syllogism

All humans are born of woman
<u>All Spaniards are humans</u>
Thus, All Spaniards are born of woman

can be represented as

All M = P   where   M = humans
<u>All S = M</u>   P = born of woman
All S = P   S = Spaniards

The false syllogism

All stones are musicians
<u>All peaches are stones</u>
So, All peaches are musicians

turns out to have exactly the same structure, where M = stones, P = musicians and S = peaches. In fact, that structure holds, as long as whatever is put for the first M is also put for the second M, whatever is put for the first S is put for the second S, and whatever is put for the first P is put for the second P--and this in detachment from how the various terms are used. We will discover in Chapter II that this particular syllogism is formally valid with respect to the process of reasoning. That the conclusion of the second example is false is not due to the process of reasoning, but to the fallacious hypotheses (premisses). If we take a syllogism,

> All men eat food
> <u>All men think</u>
> Therefore, All who eat food think

we see that two true propositions have resulted in a preposterous conclusion. Why? If we begin by asking about the first proposition ("All men eat food") the question "how many men are covered?", the answer is "all men"; so that if there are 5 billion men, "all men" means "5 billion men". But, how many food-eaters are warranted by the statement? Not all! but only as many as are needed to cover the "All men"; so that we can say that there are as many food-eaters in this proposition as there are men, and no more--which here means 5 billion. So, when we say "All who eat food" in the conclusion, we are saying more than is warranted by the premisses. This approach can be termed "syntactic" since the logical properties of the terms are decided on the basis of "co-occurrence" within a grammatical structure (once again, it could be called a "geographical" logic since where the term occurs decides its logical properties). Attention is focussed on how the terms relate not to meanings but to other terms. As basis for our study of logic, we begin with the study of sets, propositional logic, computer logic, and the relations among them.

1.1   Sets

For our present purposes, we define a *set* as "any collection of elements". The elements of which a set is composed are chosen from a universal set which contains all possible elements. We will use braces { } to indicate a set.

Examples:

$$A = \{cat,dog\} \rightarrow \text{the set of cat and dog}$$

[universal set might be "all animals"]

$B = \{2,-2,0\}$ $\rightarrow$ the set which contains 2, -2, 0
[universal set might be "all numbers of any kind"]

$C = \{a,b,c...z\}$ $\rightarrow$ the set of "all letters of the
American alphabet"
[universal set = "all letters of all alphabets"]
(note: the ... means "and so on")

$D = \{1,3,5,7...\}$ $\rightarrow$ the set of all odd numbers
[universal set could = "all numbers"]

$E = \{a,ab,ac...,abb,abc...\}$ $\rightarrow$ the set of all letter
combinations that begin with "a"
[universal set could be "all letter combinations"]

In all of the above, the universal set may be just the set itself, or any larger set.

For our later work we will want a subroutine which "inputs" sets and which "creates" sets in a nice form for output. The following is such a program:

```
1000    ! ...Set-INPUT Routine...
1100    DIM E$(100)      ! Saving 100 places for elements of the set
1110        FOR I = 1 TO 100
1120            LET E$(I)=" "
1130        NEXT I                    ! Empties set of extra elements
1200    INPUT "How many elements, less than100, are in the set ";N
1210    PRINT "Please type in ";N;"elements, with a carriage return after
            each"
1220    ! LOOP TO INPUT ELEMENTS
1230        FOR I=1 TO N
1240            INPUT E$(I)
1250        NEXT I
1260    RETURN
2000    ! ROUTINE TO CREATE SET AND PRINT IT
2010    LET SET$="{"      ! Start with the proper brace
2020        FOR I=1 TO N
2030            LET SET$=SET$+E$(I)+','   ! the Ith element, followed
                                              by a ","
2040        NEXT I
2050    IF LEN(SET$)>2 THEN SET$=LEFT$(SET$,LEN(SET$)-2)
```

```
2055    ! line 2050 removes the terminal comma
2060    LET SET$=SET$+"}"          ! put right brace on
2070    PRINT SET$
2099    RETURN
```

We wish to notice the following in the above routines:
1) LOOP 1110-1130 guarantees that no elements are in the set before we start. This is a good general practice.
2) LOOP 2020-2040 puts elements into the set, one at a time, and places a comma between the elements.
3) Line 2050 removes the comma that occurs after the last element and line 2060 puts a brace at the end.

It will be a useful exercise for us to run through the program with the set {a, b, c, ... z}. After running the first part, we would have the following values for the different variables:

E$(1)="a"
E$(2)="b"
E$(3)="c"
.
.
.
E$(13)="m"
.
.
.
E$(25)="y"
E$(26)="z"
and N=26

As we go through the second routine, SET$ is "{" at the beginning. Each time through the LOOP (2020-2040), SET$ builds as follows:

SET$={a   when I=1
SET$={a, b   when I=2
SET$={a, b, c   when I=3
.
.

```
```

SET$={a, b, c, d, e, f, g, h, i, j, k, l, m   when I=13

.
.
.

SET$={a, b, c, d, ... ,y,z   when I=26

At line 2050, the last space and comma are removed. At line 2060 the final brace is added. So, after 2060, we have:

SET$={a, b, c, d, e, f, g, h, i, j, k, l, m, n, o, p, q, r, s, t, u, v, w, x, y, z}

For any given set, we can define an important binary-valued relation: membership (or "element of"). In set A of the example before the program we find two members: cat and dog. Thus, we may say that "'cat' is a member of set A". This is abbreviated as

cat ε A

where ε means "is an element of" or "is a member of".

Each member "x" of the universe now stands in a binary relationship with set A: either *x is a member* of set A or *x is not a member* of set A. Were x a dog, we would say that x ε A is true. Were x an elephant, we would say that x ε A is false. Often, one uses the symbol ~ε to mean "not a member of". So, were x an elephant, x ~ε A would be *true*.

Thought question: Could we have chosen x to be the number "3" in the above example?

Answer: Yes and No! If our universe were all possible animals, then "3" would not be a possible choice for x. But, if our universe also contained all possible numbers (or at least the number three), which it could easily do, then we could choose x to be 3. Of course, in this case x ε A would be FALSE!

We can write a simple program which checks whether or not

an element is a member of a given set:

```
100     !...ELEMENT Program...
150     F1=0                        ! Flag for "Element in set"
200     GOSUB 1000                  ! Put elements in the set
210     INPUT "Please type in an element which you would like to check
            for membership ";X$
220         FOR I=1 TO N            ! N comes from subroutine 1000
230             IF X$=E$(I) THEN PRINT X$; "is an element of ";F1=1
                            ! indicates that the  element is in the set
240         NEXT I
250     IF F1=0 THEN PRINT X$;" is not an element of"
300     GOSUB 2000                  ! Print the set
500     STOP
1000    ! INPUT ROUTINE
1100    DIM E$(100)         ! Saving 100 places for elements of the set
1110        FOR I = 1 TO 100
1120            LET E$(I)=" "
1130        NEXT I                  ! Empties set of extra elements
1200    INPUT "How many elements, less than 100, are in the set ";N
1210    PRINT "Please type in ";N;"elements, with a carriage return after
            each"
1220    ! LOOP TO INPUT ELEMENTS
1230        FOR I=1 TO N
1240            INPUT E$(I)
1250        NEXT I
1260    RETURN
2000    ! PRINTOUT ROUTINE
2010    LET SET$="{"               ! Start with the proper brace
2020        FOR I=1 TO N
2030            LET SET$=SET$+E$(I)+','  ! the Ith element, followed
                                              by a ","
2040        NEXT I
2050    IF LEN(SET$)>2 THEN SET$=LEFT$(SET$,LEN(SET$)-2)
2055    ! line 2050 removes the terminal comma
2060    LET SET$=SET$+"}"          ! put right brace on
2070    PRINT SET$
2099    RETURN
9999    END
```

The above program on set membership works well, and uses the binary-valued relation we discussed above. That is, LOOP 220-240 compares X$ with each element of the set and if it matches an element, then a special FLAG, F1, is given a value of 1. If X$ is not in the set, the value of F1 remains 0. If we wanted, we could quicken the process by adding "I=N+1" onto statement 230. Thus,

230   IF X$=E$(I) THEN PRINT X$;" is a member of"\F1=1\I=N+1

This causes the value of I to exceed N, so the loop is exited, although in an "unnatural" way. By using this device, we leave the loop as soon as we find that X$ is an element of the set. Why check further?

Exercise: The same element is not permitted to repeat in a set - that is, {a,a} is really only {a} and really has only one element. Rewrite the set-input routine to eliminate any repetitions and reduce N to the proper number. (Be careful! You cannot alter the value of N inside a loop of which it is a counter.)

1.11   Universal Set and Null Set

In order that the concept of set make sense, we must have a universe or universal set, from which to choose our elements. This set can be purely arbitrary. Of course, we could say that the universe should be "all anythings" (whatever that means), but such is unnecessary. Were we working with numbers, why would we want "green" even to be a possible set-member? We would be quite satisfied with a universe consisting of just numbers.

We can attempt to make a definition of the universal set as follows:

U is the universal set if and only if $x \varepsilon U$ is true for every possible choice of x.

Thought question: what is wrong with the above definition?

Answer: From what are we choosing "every possible choice of x"? We must choose it from the universal set. But, that is what we are trying to define! So, we cannot choose x until we know U. So, we cannot define U until after we know what it is. But, then we have not defined it at all! (Such a definition is a circular definition--which is not a valid definition as we will see in Section 2.21). Let us, then, stick with an intuitive rather than formal understanding of universal set, as an arbitrary collection which contains all of the elements in which we may be interested.

The "opposite" of the universal set is the *null set*. The null set is the set which contains *no* elements. We generally signify the null set as 0 or { } and may formally define it as:

N is the null set if and only if x ε N is false for every possible choice of x.

Notice that we do not have the same problem here as we had in formally defining the universal set, since we have a set, from which to choose our x; namely, the universal set, no x of which is a member of N.

Thought question: Does {0} represent the null set?

Answer: No! since 0 ε {0} is true!

Thought question: How many different universal sets exist? How many different null sets are there?

Answer: There are indefinitely many possible universes; but, every null set that corresponds to each universe is indistinguishable from any other null set. Thus, there is only one null set.

*   *   *

Each of the logical connectives we mentioned earlier is describable in terms of sets, as follows:

1.12   Complementation (not)

Let us begin by making precise what we mean by the term "opposite" which we used in the previous section when we saw the null set and the universe to be opposites. Let us say that the sets A and B are "opposite", or complementary sets if

(1) for every single x ε A, we know that x ~ε B;
(2) for every single x ε B, there is x ~ε A; and
(3) every element of the universe is either in set A or in set B.

Thus, we find that A and B are opposites, in that an element of one is *definitely not* an element of the other, and they are complementary in that every element of the universe is contained in one and only one of the sets. From here on, we will consider A′ to represent the "complement of A".

Regarding complements, we should retain the following. The complement of a given set is completely dependent upon the universe which has been chosen. For example, let the universe be the set of all real numbers and let A be the set of all even integers, or A={...-4,-2,0,2,4,...}. It should be obvious that 1~ε A, so that 1 ε A′; i.e. 1 is in the complement of A. What about 1/2? Well, 1/2 ~ε A is the case and 1/2 is a real number; so 1/2 ε A′. What about cat? We know that cat ~ε A, but also cat ~ε U. Therefore, cat ~ε A′, either. As far as our universe is concerned, cat does not even exist. Of course, we could expand our universe to include the names of all animals; in which case, cat ε U and cat ε A' would hold. We can use the following binary table to define symbolically the operation of complementation:

A	A′
1	0
0	1

where   1 means "x ε A is true"
         0 means "x ε A is false"

Since membership is binary, i.e. an element either is or is not a member of a given set, then we only need the tokens 1 and 0. The first line of the table says: "given an element which is a member of A, we can then say it is not a member of A´". The second line says: "given an element which is not a member of A, we can then say it must be a member of A´". This table corresponds perfectly with the three conditions given above.

We can also represent the two sets, A and A´, pictorially:

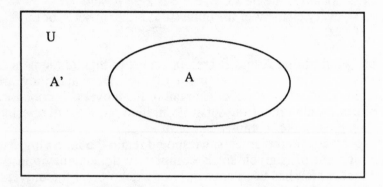

In this picture, or "Venn diagram", the rectangle defines the bounds of the entire universe. The ring (Euler ring) A defines the bounds of the set A. So, our universe is anything within the rectangle. If an element, x, is a member of the set A, it will lie within the ring A. If an element, y, is not in set A, it will lie outside of the ring A, but within the universe (for, nothing is defined outside of the universe). So, the complement of A, or A', is the set of all elements outside of the ring A, within the universe.

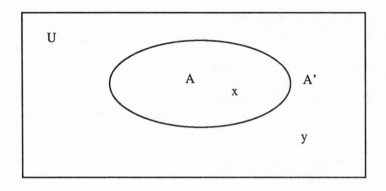

xεA
yεA'
x~εA'
y~εA

are all true statements, according to this diagram.

Thought question: Are the null set and the universal set complements, as we have defined above?
Answer: ABSOLUTELY. Since every element is in the universe and no element is in the null set, the three conditions above are trivially satisfied. Furthermore, in our symbolic table, if A = U, then the first line of the table is all that we may look at; i.e. x ε U is true for all x. Therefore, x ~ε U' for all x. But, that means that U' contains no elements.
Thus, U' = ∅.

Using our previous computer routines, we can create a program for finding the complement of a given set A in a given Universe U. The following outline will help us organize such a program:

1) We need to INPUT 2 sets instead of one--*set* A and *universe* U.
2) We need to compare each member and "eliminate" in some way the elements in A from U. In order to do this, each element of U must have a "flag", to indicate whe-

ther or not that element is in A.
3) We need to print the final set.

We satisfy these needs as follows:

1) Use the INPUT subroutine for both (with some minor alterations), but change the names of the variables, lest we lose A when we read U in.
2) Let each flag start with a value of 1. Compare each member of A to each member of U. If a member of U is in A, set the respective flag to 0, to say that the element is not in A´.
3) Create a subroutine to include only those elements with flag set to 1.

Here is our program:

```
10        !...Program COMPLEMENT...
100       DIM FS(100),ES$(100)      ! Flag and set elements
110           FOR I=1 TO 100        ! LOOP to set all flags to 1
120               LET FS(I)=1
130           NEXT I
300       LET H$="   THE SET  "
310       GOSUB 1000               ! Input the SET
320       GOSUB 2000               ! Print the SET
330       LET NS=N                 ! NS is the number of elements
                                      of the SET
340           FOR I=1 TO NS
350               LET ES$(I)=E$(I)   ! ES$ are now elements of the SET
360           NEXT I
400       LET H$=" THE UNIVERSE "
410       GOSUB 1000               ! Input the UNIVERSE
420       IF N<NS THEN PRINT "The SET cannot be larger than the
              UNIVERSE!"\GOTO 300 ! start over since sets are
                                      out of order
430       GOSUB 2000               ! Print UNIVERSE
500           FOR I=1 TO NS        ! Run through all elements of the set
510               FOR J=1 TO N     ! Run through all elements of the
                                      UNIVERSE
```

```
520                    IF ES$(I)=E$(J) THEN FS(J)=0 ! if the element of
                            the set is in the UNIVERSE, set flag to 0
530              NEXT J
540              NEXT I
550        LET H$=" THE COMPLEMENT OF THE SET "
560        GOSUB 2000                ! Print COMPLEMENT
999        STOP                      ! end of the MAIN program
1000       !...Set INPUT Routine...
1100       DIM E$(100)       ! Saving 100 places for elements of the set
1110          FOR I = 1 TO 100
1120              LET E$(I)=" "
1130          NEXT I                 ! Empties set of extra elements
1200       INPUT "How many elements, less than 100, are in ";H$;
1205       INPUT N
1210       PRINT "Please type in ";N;"elements, with a carriage return
                 after each"
1220       ! LOOP TO INPUT ELEMENTS
1230          FOR I=1 TO N
1240              INPUT E$(I)
1250          NEXT I
1260       RETURN
2000       ! ROUTINE TO CREATE SET AND PRINT IT
2010       LET SET$="{"              ! Start with the proper brace
2020          FOR I=1 TO N
2030              IF FS(I)=1 THEN SET$=SET$+E$(I)+", " ! if the flag
                         is 1, the element is in the set in question
2040          NEXT I
2050       IF LEN(SET$)>2 THEN SET$=LEFT(SET$,LEN(SET$)-2)
2060       LET SET$=SET$+"}"
2070       PRINT SET$
2099       RETURN
```

In the above program, the principal line is 520. It decides whether a member of the *universe* appears in the *set*. If it does, the flag, FS, of that element becomes 0 and this means it is not in the complement.

In the first subroutine--which starts at line 1000--the only changes from our previous version occur in lines 1200 and 1205, and these are purely cosmetic (it gives the sets their appropriate names).

In the second subroutine, only 2030 is different from what occurred before. In line 2030, an element is put into SET$ only if it does not occur in the given set, but is in the *universe*.

This program for complements can be shortened by using single line loops. That is, if a loop contains a single command--as the loop 110-130--it can be abbreviated as:

```
120     LET FS(I)=1\FOR I=1 TO 100
```

and lines 110 and 130 are deleted. Such a single line needs no NEXT statement--the NEXT is implicit in the syntax of a single line loop. Thus, the first three command lines of our shortcut program would be:

```
100     DIM FS(100), ES$(100)     ! Flag and set elements
110     LET FS(I)=1 FOR I=1 TO 100
120     LET H$=" THE SET "
```

Exercise: Rewrite Program COMPLEMENT and its subroutines, using single line loops where possible.

In line 420 we check to see whether the *set* is larger than the *universe*--which is impossible, but we never check to see if the *set* contains elements which are not in the *universe*. For example, if U={a, b, c, ..., z} and S={a, b, 2}, the set S is not in the *universe*! We solve this problem in the section on subsets and set equality.

1.13   Intersection (and)

Elements which two sets have in common are said to be in the *intersection* of the two sets. That is, given that x ε A and x ε B, we say that x is a member of the intersection of A and B. Note that the intersection of A and B is also a set, which we symbolize as A∩B or A.B. Conversely, we may say that the set A∩B contains all elements that are contained in both A and B. Our symbolic binary table is

	A	B	A∩B
line 1:	1	1	1
line 2:	1	0	0
line 3:	0	1	0
line 4:	0	0	0

The symbols, line by line, mean:

line 1:   x ε A and x ε B, so x ε A∩B
line 2:   x ε A and x ~ε B, so x ~ε A∩B
line 3:   x ~ε A and x ε B, so x ~ε A∩B
line 4:   x ~ε A and x ~ε B, so x ~ε A∩B

These are all of the possible inclusion relationships. Notice that our symbolic table is completely consistent with the definition above: i.e. only elements of both A and B are contained in A∩B and all elements of A∩B are contained in both A and B.

The Venn diagram is:

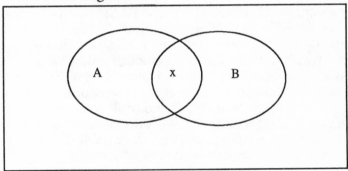

where x is a member of the intersection of A and B.

From the above, it should be clear to us why set intersection is the binary "and" relationship for sets.

Thought question: do all sets intersect?
Answer: *NO*. It is possible that they have no common element, such as the sets A = {a,b,c} and Z = {x,y,z}. In such a

case, A∩Z = 0; which is to say that the intersection of A and Z is identically the null set; and we can say that A and B are disjoint. The following definition will be useful to us.

Definition: Two sets, A and B, are identical if they have all of the same elements--element for element--in which case A=B

Our program for intersection is almost identical with the program for complementation, except that we use two sets instead of one *set* and the *universe*; and, instead of the elements which are in one but not in the other, we want the elements which are in both--which means that whatever became 0 at line 520 should be a 1 and whatever is a 1 should be a 0. Note that we must start with all of the FS's at 1 in order for line 2030 to print out the elements of the first two sets.

This change may be accomplished in one line:

545   LET FS(I)=1-FS(I)\FOR I=1 TO N

For example, IF FS(5)=0 THEN 1-FS(5)=1-0=1; on the other hand, for FS(10)=1, 1-FS(10)=1-1=0. Statement 545 is an important device for reversing the values of binary-valued functions--which is often needed in computer programming.

So that our program will work properly, the *smaller* set must be entered first, if the sets are not equal in size, or else our final printout will not go through all possible members of the intersection. That is, N must be greater than or equal to NS.

The following program finds the intersection of two sets using the abbreviated looping we discussed above.

```
10       ! ...Program INTERSECT...
100      DIM FS(100),ES$(100)
120      LET FS(I)=1 FOR I=1 TO 100
300      H$="  THE SMALLER SET "
310      GOSUB 1000
320      GOSUB 2000
330      LET NS=N
350      LET ES$(I)=E$(I)\FOR I=1 TO 100
400      LET H$="  THE LARGER SET  "
```

```
410       GOSUB 1000
420       IF N<NS THEN PRINT "You have entered the sets in the wrong
              order. Please enter the smaller one first." \ GOTO 300
,0        GOSUB 2000
,20       IF ES$(I)=E$(J) THEN FS(J)=0\FOR J=1 TO N\FOR I=1 TO NS
                                    ! notice the one line double loop
545       LET FS(I)=1-FS(I)\FOR I=1 TO N     ! reverse the 1's and 0's
550       LET H$= "  THE INTERSECTION OF THE SETS"
560       GOSUB 2000
999       STOP
1000      ! ...Set INPUT...
1100      DIM E$(100)
1120      LET E$(I)=""\FOR I=1 TO 100
1200      PRINT "HOW MANY ELEMENTS LESS THAN 100 IN ";H$;
1205      INPUT N
1210      PRINT "Please type in ";N;"elements, with a carriage return
              after each"
1240      INPUT E$(I)\FOR I=1 TO N
1260      RETURN
2000      ! ...Create and Print Routine...
2010      LET SET$="{"
2030      IF FS(I)=1 THEN SET$=SET$+E$(I)+", "\FOR I=1 TO N
2050      IF LEN(SET$)>2 THEN SET$=LEFT$(SET$,LEN(SET$)-2)
2060      LET SET$=SET$+"}"
2070      PRINT H$;" is";SET$
2099      RETURN
```

1.14   Union (or)

An element is said to be a member of the union of two sets if it is
in either one set or the other, or both. Given x ε A or x ε B, we
may say x ε A∪B; i.e. x is a member of the union of A and B.
Conversely, if x is a member of the union of A and B, it must
necessarily be a member of at least A or B. The symbolic binary
table for union is:

	A	B	A∪B
line 1:	1	1	1
line 2:	1	0	1
line 3:	0	1	1
line 4:	0	0	0

line 1:   x ε A or x ε B, so x ε A∪B;
line 2:   x ε A but x ~ε B, still x ε A∪B
line 3:   x ~ε A but x ε B, so x ε A∪B;
line 4:   x ~ε A and also x ~ε B, so x ~ε A∪B

As before, our symbolic diagram is completely consistent with our definition. In the Venn diagram

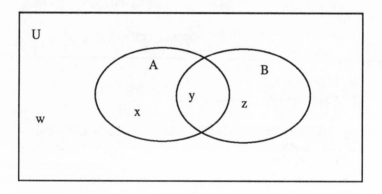

x, y, and z are all in the union of A and B, while w is not.

Thought question: can the union of two sets be empty?
Answer: Yes. But, only if both sets are empty; else the union would contain an element--namely the element which makes either of the sets "not empty".

The most difficult program to work out is one for union. If the union were merely a simple combination of two sets, the program would be quite simple. But, if the two sets had a non-empty intersection, then the elements in the intersection would be counted twice. So, in order to find the union of two sets, we must first find the intersection and then combine all of the elements of one set with all of the elements of the other set which are not in the intersection. In this case, we do not want line 545. Our union will be the entire larger set and all elements of the smaller set which are not in the intersection.

Since we allow up to 100 elements in each set, we must change the dimension statement in lines 100 and 1100 to

```
100     DIM FS(200),ES$(100)
1100    DIM E$(200)
```

Also, we must determine how many elements will be in the final set so that our loops go the proper number of times, which may not be N+NS, since N+NS counts members in the intersection twice. Our union program is identical with the intersect program, except for lines 500-550 (and line 1100, as above). We replace these lines with (in long form):

```
490     K=N
500         FOR I=1 TO NS
510             FOR J=1 TO N
520                 IF ES$(I)=E$(J) THEN FS(I)=0 ! flags the
                        smaller set
530             NEXT J
540             IF FS(I)=1 THEN K=K+1\ E$(K)=ES$(I) ! the element
                    from the smaller set will be in the union and
                    the size of the union increases by 1
550         NEXT I
553     N=K                     ! increased size of the UNION
558     FS(I)=1\FOR I=1 TO 200  ! reset flags
560     GOSUB 2000
```

1.141 A Simple Example

Let the universe be the set of integers from 1 to 10:

U = {1,2,3...10}
Let A = {1,3,5,7,9}
Let B = {2,4,6,8,10}

Let $C = \{4,5,6\}$
Then $A \cap B = \emptyset$, $A \cap C = \{5\}$, $B \cap C = \{4,6\}$
   $A \cup B = \{1,2,3...10\} = U$
   $A \cup C = \{1,3,4,5,6,7,9\}$
   $B \cup C = \{2,4,5,6,8,10\}$
   $A' = \{2,4,6,8,10\} = B$
   $B' = \{1,3,5,7,9\} = A$
   $C' = \{1,2,3,7,8,9,10\}$

Observations:
(1) Even though the element "5" is contained in both A and C, it is contained only once in $A \cup C$; That is, identical elements are *never repeated* in any set.
(2) Note that $A'=B$ and $B'=A$. It is always true that if one set is the complement of another, then the second is the complement of the first. From this we can deduce that $A''=A$ for all sets. Since $A''=(A')'=B'=A$, as per our example (in other words, the complement of the complement is the original set; double negation cancels complementation).

1.15   Subsets and Set Equality

Set A is said to be a subset of set B, that is A<B, if every member of A is also a member of B. (Note that B may have more members than A has, but it must contain at least the same members as A). More formally, we may say:

If it is true that x ε B for every single x ε A, then A<B. Conversely, if A<B and x ε A, then we may conclude that x ε B.

Thought question: Is the null set a subset of some set?
Answer: The null set is a subset of every set. Since the null set contains no elements, it satisfies the definition of subset in the most trivial way. That is, every element of $\emptyset$ (of which there are none) is an element of any other set. (Note: we will be coming back to this most difficult notion, as it will be frequently used in the sequel).

Our binary representation of the definition of A<B is:

	A	B	A<B
line 1:	1	1	1
line 2:	1	0	0
line 3:	0	1	1
line 4:	0	0	1

and requires a great deal of explanation.

Although it "looks like" the kind of table we employed to define negation, "and", and "or", there are some significant differences. The first line means "for any element x where x ε A, it is necessarily true that x ε B; from which we can conclude that A<B´´. So, the 1 in the A column stands for any element in A. In the other tables it stood for a particular element in A.

Line 2 means "for any element x where x ε A, it may be that x ~ε B, so A~<B". Another way to say the same thing is: "There exists some element x ε A such that x ~ε B, so A~<B". In other words, we only need one element for which the situation does not hold in order that A~<B. (Note that it is possible that more than one element of A is not an element of B, if A~<B).

The third and fourth lines are the strangest of all. Since 0 means x ~ε A, we are not dealing with any elements at all. As far as we can know, A is the null set. So, it makes no difference whether the element which we choose is (line 3) or is not (line 4) in B. That is, A will always be a subset of B, since it has "no" elements; for, the null set is the subset of any set. So A<B, but in the most trivial way. The following Venn diagrams represent some of the different relationships between A and B:

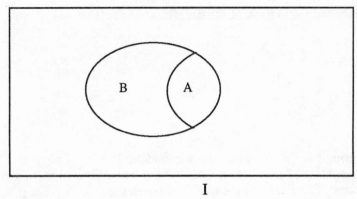

I

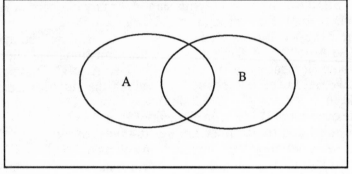

II

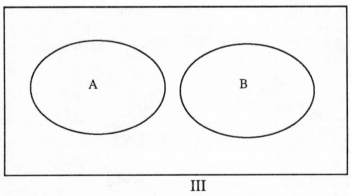

III

Diagram I represents A<B. In diagram II A~<B, although they do have some elements in common. In diagram III, A~<B and they have no elements in common. In diagram I, we see that A∩B=A and A∪B=B. In diagram II we have A∩B<>∅ (where <> means "not equal"), while in diagram III we have A∩B=∅.

If it is true that A<B and at the same time B<A, we say that A=B, that is, "A is equal to B". The only way that such can be the case is if they have exactly the same members, without exception. If A<B but A<>B, we say that A is a proper subset of B, and symbolize this as A<B. Note that A<B means that there exists at least one element z ε B such that z ~ε A.

Example:
Let A = {a,b,c}, B = {b,a,c}, C = {a,c}
A<=B since every element of A is an element of B
B<=A since every element of B is an element of A ,
so A = B (note that order within the sets does not matter)
C<=A since every element of C is an element of A
A~<C since b ε A but b ~ε C, so C < A

The following program determines whether a set is a subset of a given set. We accomplish this by using nothing more than the programming already done. In fact, our program is almost identical to the COMPLEMENT. The major change is line 520 which would be written as:

```
520     IF ES$(I)=E$(J) THEN FS(I)=0
```

The only change in this line is FS(I) instead of FS(J). In the COMPLEMENT program, the flag FS followed elements of the larger set (universal) so FS(J)=0 flagged the Jth element of the *universe*--indicating that that element was also in the *set*, and so not in the complement. Now, FS flags the subset, so FS(I)=0 indicates that the member of the assumed subset is a member of the larger set. After we have gone through all of the members of the subset, FS(I) should be 0 for every I--else such ES$(I) is not a member of the larger set, and so the assumed subset is not a subset. To complete the program, we must check that all the flags are 0. Thus:

550   IF FS(I)=1 THEN NO=1\FOR I=1 TO NS
560   IF NO=1 THEN PRINT "The assumed subset is not a subset since
                the following are not members of the larger set"
                \ELSE PRINT "The subset is correct"
570   IF FS(I)=1 THEN PRINT ES$(I)\FOR I=1 TO NS

The above three statements do the following:

    550   If some element is not in the larger set, NO=1
                    indicates that the set is not a subset.
    560   prints out whether a set truly is a subset.
    570   prints out any members of the smaller subset which
                    are not in the larger set.

The other changes which must be made are trivial, such as changing universe to "larger set" and "subset" to "assumed subset".

Exercises 1 (Programming):
1) Write a subset program.
2) Write a subset program as a subroutine which can be incorporated into the COMPLEMENT program to check to see that the given set is truly within the prescribed universe.
3) The difference, A-B, of two sets is defined as the set containing all elements of A which are not in B.
For example, if A={a, b, c, d} and B={b, d, f} then
    A-B={a, c} and B-A={f}.
Write a program which finds A-B and B-A.

Exercises 2: Let A = {a}, B = {a,b}, C = {a,b,c}, ... Z = {a,b,c,...,z}, where the universe is the lower case letters of the alphabet:
    i) Find A´, P´, Z´
    ii) Find A∩B, A∪B, M∩P, M∪P
    iii) Find the set {b} as union, intersection, complement of sets from the above.

iv) Find the set {b,d,m,z} from the sets above (as in iii).

1.2   Propositional Logic

1.21   Truth Tables

Another meaning for "and", "or", "not", "is", "if...then..." can be found in propositional logic, where we study the truth of propositions and the validity of certain combinations of propositions. But, what is a proposition?

> *Definition*: A proposition is a statement which is either true or false. It cannot be both true and false at the same time and in the same respect, but must be true or false at any given time in a prescribed way.

The simplest example of a proposition would be something like "it is raining". Such an assertion is always either true or false, for it cannot be both raining and not raining in a particular place at a particular time. Notice that this statement may be true in Boston at 10 a.m. on Friday the 13th, but false in Geneva at the same time. Or the opposite could be the case. Or, the proposition might be true at that time in Boston and false in Boston at 10:01. Thus, we understand a proposition at a particular time taken in a particular way (the "way" here is the location).

Not all statements are propositions. There are, for example, questions, commands, exclamations, paradoxes and self-contradictory statements. What is more, the same statement may be a proposition when taken one way and not a proposition when taken in another way. For example,

    (1)   I am a fool!
    (2)   I am a fool?
    (3)   I am a fool.

The first two statements are not propositions while the third is.

Another example of a statement that is not a proposition is the paradox (Greek: *para* + *doxa* = against belief). Self-contradictory statements or paradoxes are true when taken to be false and vice versa. For example, "This statement is false" is self-contradictory. When it is true, it is false; and, when it is false, it is true. Such statements are the subject of advanced logical analysis and will not occupy us here (see Russell's Paradox in Section 1.5).

1.22   Negation

A simple or atomic proposition is one which cannot be broken down into smaller parts while still remaining a proposition. The negation of such a proposition is straight-forward--the negation of a true proposition is false and the negation of a false proposition is true. If we take p to represent a proposition (e.g. p = "it is raining") and ~p to represent the negation of p (~p = "it is not raining"), we can set up the following table:

p	~p
T	F
F	T

where F represents false and T stands for true.   This representation is called a "truth-table", for it represents all possible truth values (T or F) of both p and ~p. This particular truth-table is the definition of negation. Note well that when p is a proposition, ~p is also a proposition.

1.23   Conjunction (and)

Atomic propositions (i.e. propositions which cannot be reduced to simpler propositions) can be combined in various ways, which produce compound or "molecular" propositions.   These combinations can also be defined by truth tables. For example, the truth-table representation for conjunction (and) is:

p	q	p•q
T	T	T
T	F	F
F	T	F
F	F	F

What this table represents is the following:

The conjunction p•q is true when p and q are both true propositions. Otherwise, p•q is false. That is to say p•q is true when both p and q are true. For example, let p = "it is raining" and q = "the grass is wet". Now, p•q represents "it is raining and the grass is wet", which is true only when it is both raining and the grass is wet! Note that this conjunction, p•q, is a proposition--it is either true or false. Its truth depends upon the truth of each of the atomic propositions which make up the conjunction.

We thus have the following two principles:

(1) The conjunction of any two propositions is always a proposition.
(2) The conjunction of two propositions is true when both of the propositions are true, and false under any other circumstances.

1.24   Disjunction (or)

The "or" function of propositions is called disjunction and is represented by the following table:

p	q	pVq
T	T	T
T	F	T
F	T	T
F	F	F

If p = "it is raining" and q = "the grass is wet", then pVq = "it is raining or the grass is wet". The definition of disjunction reveals the following:

(1) pVq is a proposition; i.e., the disjunction of two or more pro-

positions is a proposition.

(2) The disjunction pVq is true when either p or q is true or both are true. It is false only when both p and q are false.

1.25 Implication (if...then)

The most frequently used and also most complicated of the truth functors in modern logic is the material implication that takes the form of the conditional proposition. The truth-table for material implication is:

$$
\begin{array}{ccc}
p & q & p{\rightarrow}q \\
\hline
T & T & T \\
T & F & F \\
F & T & T \\
F & F & T
\end{array}
$$

The $p{\rightarrow}q$ can be read in many ways, including:

p implies q
if p then q
p is sufficient for q
q is necessary for p
p only if q

In the material implication, $p{\rightarrow}q$, we call p the "antecedent" or premiss and q the "consequent" or conclusion. Unlike conjunction and disjunction, where q•p has the same truth value as p•q and qVp is truth-table equivalent to pVq, $q{\rightarrow}p$ does not have the same truth value as $p{\rightarrow}q$. Implication is "precedence determined" (it matters which proposition comes first). That is, implication does not commute, while conjunction and disjunction do commute. We may say that:

(1) The material implication formed by two propositions is a proposition.

(2) The material implication is false only when a true antecedent leads to a false consequent. It is true when a true antecedent

leads to a true conclusion or when a false antecedent leads to
any consequent.

How are we to understand paragraph (2)? Our best insight
into the problem of implication is through our earlier comments
on sets. Note that the table for set intersection is exactly the same
as that of logical conjunction (with 1 for T and 0 for F). Such is
also the case for union and disjunction, and for complementation
and negation, as well as inclusion and implication.

Although set theory and propositional calculus are different
fields (since a proposition is not a set), there are similarities
between the two so that each helps us to understand the other.
By changing the language of sets to the language of propositions
we can interpret p→q as "p 'is a subset of' q"; that is, the truth of
p must guarantee the truth of q. So, when p is true and q is true,
we have no problem. When p is true and q is false, something is
wrong, since the truth of p is supposed to guarantee the truth of q
as "an element in p must also be in q"--but, it does not do so.
So, the entire proposition p→q is false.

A false premiss implies any conclusion since it guarantees
nothing whatsoever about the conclusion. *E falso sequitur
quodlibet* (from the false anything follows).

Let us return to the implication, where p = "it is raining" and
q = "the grass is wet". p→q =then means "if it is raining, then
the grass is wet". That it is raining is sufficient to guarantee the
grass being wet. Similarly, wet grass necessarily results from
rain. So, rain (p is true) and wet grass (q is true) tell us that p→q
is true. What if it rains (p is true) but the grass remains dry (q is
false)? Then the proposition p→q must be false, since rain is a
condition which guarantees wet grass. But, it rains and there is
no wet grass! So, p→q is false. But, what if it is not raining (p
is false)? The grass may or may not be wet (the sprinkler may
have been on to wet the grass--another sufficient condition to
guarantee wet grass). It makes no difference whether or not the
grass is wet since only a true premiss guarantees the conclusion.
A false premiss guarantees nothing whatsoever; it is as without
content as the null set is without members. This means that any
conclusion "works" and the implication p→q is trivially true. "If
a number is divisible by 7 but not by 2, it is odd" illustrates how

a conditional sentence is true under various conditions. "Twenty-one" fulfills the condition and the consequent obtains. "Twenty" is a number for which the consequent does not obtain because the condition is not fulfilled. The statement is compatible with the fact that although "nine" does not fulfill the condition it is odd. What is excluded, obviously, is that any even number can fulfill the condition. Similarly, "If anyone is a committee chairman in the Massachussetts Legislature, he or she is a Democrat" remains true (possibly until the next election), although there are hundreds of thousands of Massachussetts residents for whom the antecedent is not true, but the consequent is.

Although it may be upsetting to many that a false premiss implies any conclusion--just as the null set is a subset of any set--it should be remembered that such truth is not terribly meaningful. Only in logic do we concern ourselves with such "trivial" truth. In "real life" we always try to avoid such trivialities. But in logic, such trivial matters can prove instructive and even important.

1.26   Biconditionality (Equivalence)

The final truth functor we will consider here is the biconditional which is also called equality or equivalence. Its truth-table is:

p	q	p↔q
T	T	T
T	F	F
F	T	F
F	F	T

In other words, p↔q is true only when both p and q have the same truth value. Notice that the biconditional formed by two propositions is also a proposition. We may speak of p↔q as:

p and q imply each other
p is necessary and sufficient for q
p if and only if q

The biconditional is thought of as the "is" or equality in propositional logic, since a true biconditional ensures that the propositions on either side have the same truth value. Thus, one can substitute the proposition p for the proposition q without altering the truth value of a given functor.

Thought question: is (p•q)→(pVq) a proposition?
Answer: Certainly. As long as both p and q are propositions, then both p•q and pVq are also propositions and their combination in (p•q)→(pVq) is also a proposition. In other words, atomic or elementary propositions can be used to form compound or molecular propositions.

Exercises: Some of the following combinations of symbols are propositions. Letting p = dogs bark, q = cats meow, and r = frogs croak, write the propositions in proper English.

   a) pV~p
   b) ~(q•~q)
   c) p→(q•r)
   d) (pVq)↔(qVp)
   e) p→q→r~p
   f) p→(q→p)
   g) (p•q)→(pVq)
   h) ~Vq→rV(p•pr)

1.3   Quantifiers

Having seen the basic notions of set theory and of truth-table verification, we can now discuss how the number of elements in a set satisfies certain conditions. Actually, our discussion is concerned with whether or not all the members of a set satisfy a given condition, or whether some members satisfy such a condition.

Let x be a variable chosen from a particular set S. Let Fx be a condition F on x (a condition becomes a proposition when a value is selected for the variable x). For example, if S is the set of all humans, then x is a human chosen from S; Fx could be the condition "x is rational". When x = John, "John is rational" is a proposition which, of course, is either true or false.

> *Definition*: The universal quantifier asserts that a condition is a true proposition for every member of a given set. The universal quantifier is symbolized by (x), which is read as "for all x", "for each x", or "for every x". The existential quantifier asserts that there exists at least one member of set S, for which the condition is true. The existential quantifier is symbolized as (∃x) and is read as "there exists an x" or "for some x".

Continuing our example, the statement (x)Fx means "all humans are rational" while (∃x)Fx means "there exists a human who is rational". We find both of these sentences to be true. Notice that (∃x) does not exclude the possibility of "more than one" or, in fact, the possibility of "all". (∃x) asserts merely that at least one exists!

From the same set, the set of humans, let Gx mean "x is over seven feet tall". In this case, (x)Gx, "every human is over seven feet tall" is false, while (∃x)Gx, "there is a human who is over seven feet tall" is true. When Hx means "x has 3 eyes", we find both (x)Hx, "all humans have 3 eyes", and (∃x)Hx, "there exists a three-eyed human", are false.

These examples have provided us with cases where the statements (x) and (∃x) were both true for a given function, both false for another function, and (∃x) was true and (x) false for a third function. Is it possible for (x) to be true while (∃x) is false?

Although it seems unlikely, it is possible that (x) be true and (∃x) false, but this can only occur when the set in question is the null set. Why is this the case?

The universal quantifier asserts that a function holds for every member of a given set. But, it never states that there are in fact any members of that given set. The existential quantifier, however, asserts that an element must exist for which the function is true; so that a function cannot be true when no element exists.

An instance of such a case is the set of unicorns, which is

empty. Paradoxically, some would say that for Fx as "x has a horn", the statement (x)Fx "all unicorns have a horn" is true while (∃x)Fx, "there is a unicorn which has a horn", is false, since there are no unicorns. Once again, the null set makes its strange appearance.

One wonders why we only have two quantifiers. Why is there no quantifier for "none" or for "some are not" ? As it turns out, both "none" and "some are not" can be expressed with the help of (x), (∃x) and the operator of negation. Simply stated, "none" is exactly the same as "all are not". Returning to our first example, (x)~Fx would mean "all humans are not rational" or "no human is rational", while (∃x)~Fx means "some human is not rational".

1.31   The Negation of Quantifiers

This brings us to the thorny question of the negation of quantifiers. For, after all, what do ~(x) ("not all") and ~(∃x) ("not some") mean? The first guess might be that ~(x) is the same as "none". But, this is not the case.

As with any negation, if (x)Gx is true, ~(x)Gx must be false. Conversely, when (x)Gx is false, ~((x)Gx) must be true. Such is the case with (∃x)Gx and ~(∃x)Gx. In our second example, where x is the set of humans and Gx = "x is over seven feet tall", we found (x)Gx to be false. Thus, ~((x)Gx) must be true, so it cannot mean "No humans are over seven feet tall", for this statement is also false. "No human is over seven feet tall" is represented as (x)~Gx.

It turns out that the only possible negation of "all humans are over seven feet tall" is "there exists a human being who is not over seven feet tall"--which we express symbolically as (∃x)~Gx. So, we find the negation of "all" to be "some are not": ~(x)Gx=(∃x)~Gx. Similarly, the negation of the true statement (∃x)Gx, "there is a person over seven feet tall", must be a false statement--"no people are over seven feet tall" or "all people are not over seven feet tall". Therefore, ~∃xGx=(x)~Gx.

Note that the negation of a negation is the original sentence: ~((∃x)~Fx) = (x)~~Fx = (x)Fx, since ~~Fx=Fx.

With the above in mind, we are in a better position to interpret our unicorn problem. Since there are no unicorns at all, why is (x)Fx, "all unicorns have horns", true? (x)Fx is true because its negation, ~(x)Fx = (∃x)~Fx, "there is a unicorn which does not have a horn", is false, since (∃x)~Fx asserts the existence of a unicorn (who cares about its "horned" condition). Note that (x)~Fx, "all unicorns do not have horns" is also true, since ~((x)~Fx) = (∃x)~~Fx = (∃x)Fx = "there is a unicorn with a horn", is false since there are no unicorns. What we see here is that any condition whatsoever is true when asserted by or of the null set, just as any conclusion is "justifiable" by a false premiss. In a null set, the negation of "all" is not "none". So, beware of the null set! There is a profound sense in which "our shadows have no faces".

By understanding the above negation, we can show that what was once considered the "paradox of the liar" is no paradox at all. It is maintained that Epimenides the Cretan said "All Cretans are liars". He, being Cretan, must have been lying. So, it is false that "All Cretans are liars". Were one to think the negation of "all" was "none" then we would have a problem, since "All Cretans are liars" and "No Cretans are liars" indeed cannot be simultaneously true but they can be false at the same time. But, since the negation is "Some Cretans are not liars" we have a statement that could be true, leaving the possibility that some Cretans are liars--in particular Epimenides himself! So, not only is "Epimenides says 'All Cretans are liars'" not a paradox; but it is a statement from which we can draw the valid conclusions that

> Some Cretans do not lie
> Epimenides is a liar

A final note on notation: When a variable is quantified, we call it a "bound variable". Otherwise, it is a "free variable". The variable x in "x > 5" is free, while x in (x) x > 5 is bound, as is x in (∃x) x > 5.

Exercises:

1. If Fx represents "x is mortal" and Gx represents "x is a man", write the following in proper English:

A) (x)Fx
B) (∃x)Gx
C) ~(x)Fx
D) (∃x)(y)Fx→Gy
E) (y)(∃x)Fx→Gy
F) (∃x)~Gx•Fx
G) (∃x)(∃y)Fx→Fy
H) (∃x)(y)Fx→(FxVGy)

2. Let Lxy represent the binary statement "x loves y". Then (x)(∃y)Lxy would mean "everyone loves someone", while (∃y)(x)Lxy means "someone is loved by everyone". Please interpret the following:

A) (x)(y)Lxy
B) (∃x)(y)Lxy
C) (∃x)(∃y)Lxy
D) (y)(x)Lxy
E) (y)(∃x)Lxy
F) (∃y)(∃x)Lxy

1.4   Computer Logic

Thus far we have seen two examples of two-valued (or binary) systems: the algebra of sets (Section 1.1), and the calculus of propositions (Section 1.2). We have also seen quantifiers as a link between sets and logic. The logic of the computer is analogous to what we have done so far.

Binary logic is well suited for the computer, since our ordinary digital computer is merely a collection of electrical circuits which are either *on* or *off* (magnetized or not, open or closed, charged or discharged, etc.). All information in such a computer is stored in on-off circuits, and all computer processes and calculations are achieved by switching the proper circuits to "on" or "off". Of course, there are other technical concerns, like getting information into and out of the computer, but these need not concern us here. Our concern is with the structures involved in storing and using information in the computer.

A piece of information is stored in a computer component which is either *on* (represented by 1) or *off* (represented by 0). A single such component-state is called a "bit" and a collection of bits is a "byte" (usually 8 bits). A collection of bytes is called a "word" (usually 1, 2 or 4 bytes to a word). Thus, the language of the machine is a language of 1's and 0's (ONs and OFFs). We now need to know how a piece of information actually is stored.

The simplest information to be stored are positive integers. These are stored in words using binary (base two) notation as follows:

integer	byte
1	0 0 0 0 0 0 0 1
2	0 0 0 0 0 0 1 0
3	0 0 0 0 0 0 1 1
4	0 0 0 0 0 1 0 0
5	0 0 0 0 0 1 0 1
6	0 0 0 0 0 1 1 0
7	0 0 0 0 0 1 1 1
107	0 1 1 0 1 0 1 1

In this binary notation, each place represents a power of two, just as in decimal notation each place represents a power of 10. ($10{\char`\^}2$ in BASIC means what 10^2 means in mathematical symbolism)

10 000	1000	100	10	1		128	64	32	16	8	4	2	1
$10{\char`\^}4$	$10{\char`\^}3$	$10{\char`\^}2$	10	$10{\char`\^}0$		$2{\char`\^}7$	$2{\char`\^}6$	$2{\char`\^}5$	$2{\char`\^}4$	$2{\char`\^}3$	$2{\char`\^}2$	$2{\char`\^}1$	$2{\char`\^}0$

Just as the number 7862 represents 7 thousands, plus 8 hundreds, plus 6 tens, plus 2 ones

$$(7(10{\char`\^}3)+8(10{\char`\^}2)+6(10{\char`\^}1)+2(10{\char`\^}0)),$$

the binary number 101101 represents one 32, plus one 8, plus one 4, plus one 1

$$(1(2^5)+ 1(2^3)+1(2^2)+1)$$

which would be 45 in decimal notation. It is important to note that any number (integer, rational, irrational, transcendental) which can be represented in base 10 (decimal notation) can be represented in base 2 (binary notation). We just need more "places" in base 2 than we do in base 10.

In order to represent a decimal in the computer, we need more bytes. Two bytes tell us what the value of the number is without the decimal point, and another one or two bytes tells us where the decimal point goes. Negatives are represented as the inverse of positives, and letters are converted to numerics and then represented in the same way as positive integers. For example, in the ASCII coding system (American Standard Code for Information Interchange) the letter "A" is converted to the number 65 and stored as 1000001. Of course, there is a special "flag" which lets the computer "know" that 1000001 is the letter "A" and not the number "65". The details of this encoding are beyond the range of our interest here. What we want to retain is that whatever can be represented with letters and numbers can be converted to binary notation and stored in the computer--provided the computer has enough memory (i.e. storage space). Arithmetic operations (addition, subtraction, multiplication, division, exponentiation, squaring) can be done even more easily in binary than in decimal. For example, $181 + 230 = 411$ comes out as $10110101 + 11100110 = 110011011$, as follows:

256	128	64	32	16	8	4	2	1	
	1	0	1	1	0	1	0	1	181
+	1	1	1	0	0	1	1	0	+ 230
1	1	0	0	1	1	0	1	1	411

[notice in the 4's column, $1 + 1 = 10$; so, 1 is "carried". Note also the "carries" in the 32, 64 and 128 columns]

As numbers and letters can be represented and manipulated through circuits within the computer, so too can the functions of propositional logic. Through the miracles of electrical engineering with semi-conductors, one can create logical circuits or "gates" to represent the logical functions of propositional calculus. The simplest of these are the "not" gate, the "and" gate, and the "or" gate, which we represent as:

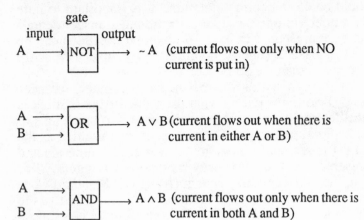

These gates can be connected together in various ways to represent any logical function. For example:

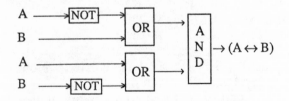

represents the implication, while

A ——→[NOT]—→ ⌐OR⌐ ——→ ⌐AND⌐ → (A ↔ B)
B ——————→
A ——————→
B ——→[NOT]—→

represents the biconditional.

The validity of the above representations of implication and the biconditional are proven in Section 2.351.

1.5   Russell's Paradox

Bertrand Russell (1872-1970), a famous modern British logician, discovered that a paradox results with respect to the definition "a set is any collection of elements". His discovery was based on the following.

Since a set is "any collection", let us consider two sets:

A = "the set of all sets which do not contain themselves as elements", and

B = "the set of all sets which contain themselves as elements"

It should first be noted that most sets we naturally conceive do not contain themselves as elements. For example, C = {1,2,3} does not contain itself as an element, else it would be D = {1,2,3, {1,2,3}}. Now, D contains C, but D does not contain itself as an element! It seems that no set could contain itself as an element. But, what about the following: P = "the set containing all sets describable in less than 12 words". Now P contains itself as an element, since the set P is described with 11 words. P may seem bizarre as a set, but so is the condition upon which it is based, i.e. a set which contains itself as an element. (What if p = "the list of all books in this library", where p is a book in the library, too?)

Now, let us consider the set A, which is {all sets which do not contain themselves as elements}. Does A contain A as an element, i.e. A ε A ? Well, if A contains itself as an element then it is a set which does not contain itself as an element, so A ε A → A ~ε A -- which is ridiculous. On the other hand, if A does not contain itself as an element, it must be a member of itself, since A is the set of all sets which do not contain themselves. So, A ~ε A → A ε A -- which is absurd. Thus, A must contain itself as an element, but it cannot contain itself as an element--which is going from the ridiculous to the absurd. Consequently, something must be wrong with the definition of set as "any collection", for we have a collection which logically cannot be a set.

Recreating a definition of a set in order to eliminate the

Russell Paradox has been an arduous task, taken up by many eminent logicians, including Frankel, Skolem and Zermelo. Books on formal set theory contain further information on this subject.

Problem: If the barber is the person who shaves everyone who does not shave himself, who shaves the barber?

II
Elements and Manipulations

2.1   Terms

Logic, we know, studies the laws of the manipulation of valid
forms, including those of thought. To understand the elements of
logical manipulations (also called "reasonings") it is necessary to
develop a kind of grammar--a special, logical grammar with a
limited number of categories; for, the tangible expression of our
thought is to be found in written and spoken language. Although
it is often easier to refer to language rather than to thought, it is
worth noting that logicians in principle refer to thoughts (or
ideas) rather than to words, when they talk about propositions
and terms. Thus, a proposition about what we judge to be true or
false is usually expressed by a sentence, and the elements or
terms of that proposition--the concepts that are joined together in
our judgement--are expressed by words or groups of words.
   Although some contemporary logicians have studied the logic
of commands and of questions, work today--as in the
past--concentrates on propositions (judgements) that are
expressed by declarative sentences. One might say that words
signify terms or concepts which, in turn, signify things, and that
sentences signify propositions or judgements which represent
states of affairs. Concepts are a unique kind of sign, because we
understand what they signify without needing to be told. There
is the observation in Lewis Carroll's *The Hunting of the Snark*:

> "What's the good of Mercator's North Poles and Equators,
> Tropics, Zones, and Meridian Lines?"
> So the Bellman would cry and the crew would reply
> "They are merely conventional signs!"

Of course, the good logician Lewis Carroll is playing with the
difference between something conventional and something purely

arbitrary and meaningless. In fact, words, letters (for sounds), or a striped pole for a barber shop are conventional. Concepts, on the other hand, are natural signs. However, it is not enough to say that terms stand for things. Apart from obvious examples like "nothing" which St. Augustine puts to Adeodatus in the *De Magistro* to break down the simplistic concept-sign-for-thing equation, there is an important class of signs that represent logical relationships and operators rather than things. This is related to the problem of "Nothing is better than peanut butter". Does this say peanut butter is good or bad? Are we saying "I would rather have nothing than peanut butter" or "There is nothing which is better than peanut butter"? Such is the problem with having nothing.

Hence, we will distinguish between *categorematic* signs and *syncategorematic* signs. Roughly speaking, the categorematic signs represent things and their properties--or, at least, things like "gobblins", "hobbits", and "snarks" that are thought of as things--while the syncategorematic signs include "if...then...", "not", "all", "some", "or" and also "is". Logic studies syncategorematic terms and often replaces the categorematic terms by letters, numbers, or other symbols, so that one is not distracted by interesting content from the study of logical form. We have examined this symbol substitution in Chapter I, above. By and large, syncategorematic terms serve as functors or operators and categorematic terms as variables. The recently popularized PROLOG computer language constructs procedures with precisely this distinction in mind:

```
egg(poached).          /* begin database */
egg(hard-boiled).
egg(scrambled).        /* note the terminal periods */
egg(fried).            /* note the lower-case */
egg('Easter').         /* see how these comments are formed */
egg(sunny-side-up).
egg(three-minute).     /* end database */

egg(X).                /* THE rule */
```

Here X is a categorematic sign which stands for the different possible kinds of eggs. We may say X ε {poached, hard-boiled,

...}. "Egg" is a syncategorematic sign or functor.

```
cooks('Sally',eggs).      /* database begins */
cooks('John',bacon).      /* note single quotes used so that caps are */
cooks('Henry',toast).     /* not confused with names for variables */
cooks(he,breakfast).      /* database ends */

cooks(X,Y).               /* the rule */
```

In this example we have two categorematic signs, X and Y, where X is a variable from {Sally, John, Henry, he}; Y is a variable from {eggs, bacon, toast, breakfast} and "cooks" is the syncategorematic functor which provides a relationship between X and Y.

Notice that we cannot write

```
cooks(sally,john).
```

since john is not a member of the set from which we may choose Y.

However, it is worth noting that when we ask what a categorematic term means, we can answer that question in two ways--by explaining what thing it means, or by indicating what ideas it suggests. Logicians have several ways of expressing this distinction: the things a term means are its *extension*, denotation, or reference; the idea the term suggests are its *comprehension* (also "intension"), connotation, or sense. Thus, to express the same distinction, there are three (actually three and a half) pairs of terms: extension-comprehension (or extension-intension), denotation-connotation, reference-sense. Students having trouble in their beginning philosophy course may find it consoling to note that full-grown philosophers frequently do not understand each other's language, either. To make matters worse, the student may be confronted by a book in English composition which uses the doublet denotation-connotation in a different way: now, denotation becomes what a term objectively means (what the logician calls "connotation"), while "connotation" becomes its personal, subjective overtones. Thus, New York City might have a connotation of boredom for someone whose experience has been limited exclusively to changing planes at Kennedy Airport, while "beef stew" might suggest the grandmother who excelled at preparing that dish. Philosophers in general eschew such subjective entanglements and hence concern themselves with the two objective ways of signifying.

Exercises:
1. List part of the extension (denotation) of "human".
2. Is "rational animal" the comprehension (connotation) or extension (denotation) of "human"? Explain briefly why.
3. Does "person who became king of France in 1987" have a sense? a reference? Explain briefly.

2.11   Definition

Either way of signifying--reference (extension, denotation) or sense (comprehension, connotation)--may be of help in explaining precisely what a term means, in giving a definition. The nature of the definition may be highlighted by saying that the copula "is" in a definition has critical importance. It is true that "The cow is an animal that gives milk" but this is not a definition of cow, because the trait is not limited to cows. To define cow, we want to give a description that applies to all cows and only to cows, that denotes the same individuals, where the definer (*definiens*) is interchangeable with what is to be defined (*definiendum*). To have a true definition, we must offer a description that is neither too broad--applying to more things than it should, nor too narrow--not including things that it should.

Furthermore, we want a description that says something important about what we are defining, that captures its essence, or pinpoints a cause that makes it be the way it is. "Man is a featherless biped" fails to do this, not because there are other featherless bipeds, nor because some people have feathers, but because, unlike "Man is a rational animal", the first statement gives us superficial or accidental characteristics.

There are a variety of ways in which a statement can be especially defective in telling us what the definiendum is like.

Negative descriptions are indefinite. Even if we manage to exclude all but the group we wish to define (and negative definitions, by their indefiniteness, lend themselves to broadness), we do not give a clear idea of what our definiendum is in fact. It would not help much to say that "A plant is what is neither animal nor mineral". Of course, there are some concepts, like "bachelor" or "blind", which are inherently negative, meaning "not married" and "lacking sight", so that the definition has to stress this negation.

Similarly, our definition should not be figurative or metaphorical. It would not be adequate to say that "The lion is the king of beasts", because this makes sense only when we already understand what a lion is.

The same point could be made even more emphatically about circular definitions--ones that include the definiendum or a close derivative in the delimiting statement: "A twin is someone who has a twin brother or sister."

Since a definition is the basis for understanding and discussion, one should not try to settle arguments ahead of time by sneaking one's interpretations, conclusions, or value judgments across as definitions. Thus, it may be that "logic is the basis of all study of philosophy", but that is a conclusion that can be appropriately discussed only after we know analytically what logic is. To be sure, some terms are inherently value-laden, just as some terms are negative. One cannot explain what "murder" is, distinguishing it from the more generic "homicide", unless one includes some notion of culpability.

We now have seven rules: definitions should *not* be (1-6):
1. too broad,
2. too narrow,
3. negative,
4. figurative,
5. circular, or
6. problematic interpretations, but
7. they should say something important.

A bad definition--which might better be called "mere description" --will often violate more than one rule. It will be noted that--except for the seventh point and, perhaps, the first two which can be conflated as "interchangeable" or "coextensive"--most of these rules are negative and do not help the person who attempts to construct rather than just to criticize a definition.

One helpful suggestion is contained in the traditional notion of definition by genus and specific difference. One first finds the larger group into which something fits (but not too large a group), and then finds the characteristic or group of characteristics which set the thing apart from other members of this same larger group (the genus). Thus, the genus of "man" is "animal" and the specific difference is "rational". Or, the genus of "chair" is "piece of furniture" and the specific difference might be "with a back that seats one person".

The genus\specific difference model may not always work --especially with artificial objects (e.g. an automobile)--but it provides a helpful pattern that can be worked out in two (or three or four, if the specific difference is elusive) steps.

In an older tradition there was the *Tree of Porphyry*:

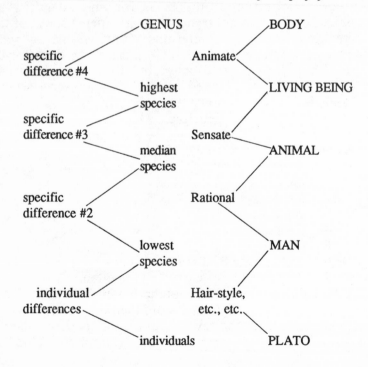

This tree can still serve the purpose of showing how a good definition is structured, and how division (see below) can be put to use. Anything that falls under *genus* in the diagram can serve as subset of whatever is higher in the same column, from which it is distinguished by whatever occurs in the correlative position in the left column.

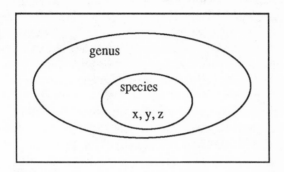

For example, if the highest *genus* is "means of transport", an adequate division would be between "motor-driven" and "other-powered" (whereas "horse-drawn" and "ox-drawn" would be an incomplete division). The specific difference here would be "motorized". Further, "motor-driven" could be divided into "by internal combustion" and "by external heat" (whereas "wood-burning" and "coal-burning" would, again, be incomplete). We could continue the process until we arrive at a definition of automobile as a "means of transport driven by an internal combustion engine". The person who works out definitions in this manner will also grasp the general relation between extension and intension: as the sense of a term becomes more complex, it refers to fewer things, and vice versa. For example, the set of red chairs is equal to or smaller than the set of chairs; while the set of three-legged red chairs is even smaller. In each case, as we add more traits (properties), fewer items are included.

Most of what we have said about definitions involves making clear the connotation of a term. One could pinpoint a definition denotatively, e.g. explaining "President of the United States" by giving a list (denoting those who fit the concept to be defined). Often, the quickest way to communicate is to point out an example, i.e. to indicate a "love-seat" at the furniture store or to show an act of kindness. (But showing an act of kindness on a love-seat may or may not help to define "love-seat"!) This practice is known as "ostensive" definition. If something is an outstanding example of its type--say Douglas MacArthur as "military hero"--it can become a "paradigm". However, the use of ostensive definition or of paradigm cases violates the cardinal rule that our definiens must be coextensive with the definiendum, since in these cases the *definiens* would be too narrow.

Exercise 1: Find the most obvious rule violation (there may be more than one, but look for the flagrant one) in each of the following definitions and briefly say what precisely the problem is.

Sample:

Whales are not land animals, nor fish, nor invertebrates, nor dolphins, nor sea-lions, nor seals.

This is negative (it is probably too broad); it only says what whales are *not*.

1. A man is like John.
2. Humans and angels sing in choirs.
3. The designated hitter hits in place of the pitcher, sits on the bench until then, and goes into action only at the behest of the manager.
4. An animal does not use articulate speech, does not argue in syllogisms, and cannot drive a car.
5. A hit is the impacting by the batter of the cowhide-covered spheroid.
6. Socialism is democracy in the economic sphere.
7. Cats are pets that meow.
8. College professors are people who profess some body of truth at a college.
9. Physicians are highly trained professionals.
10. Physicians are men and women who perform surgery on humans.
11. Tigers are striped mammals.

12. Life is a bowl of cherries.
13. Life is a sewer; you get out of it what you put into it. (T. Lehrer).
14. Mineral is what is neither animal nor vegetable.
15. Dogs are beagles and collies.

Exercises 2:
1. Explain briefly the connotation of "President of the United States". What is its denotation?
2. Which of the following are synonyms of "extension": connotation, denotation, intension, sense, reference?
3. Arrange the following in order of decreasing extension: printed matter, volume of religious significance, Bible, book, material object.
4. Which of the following are propositions?
   a. Have you ever seen so sad a tiger?
   b. Tigers are purple.
   c. Kill the tiger.
   d. Tigers have tails.
   e. Tigers, wow!
5. Pick out the syncategorematic terms in the following list: pig, chicken, if, green, because, truck, therefore, marble, not, implies, love, sour, inasmuchas.

2.12   Classification and Division

Another way of increasing our understanding is to classify. This involves dividing up a group and organizing its extension or reference. Classification is different from definition. Value judgements are involved in classifying; for example, professors grade students (and vice versa), and it is at this point that practical considerations intervene: in the United Nations library, the languages in which books are written might be a useful starting point for splitting them up. In most American high school libraries it certainly would not be useful; for, the integrating parts to such a division would be too lopsided, most of the books being in English.

To be sure, lopsidedness itself is not inevitably a problem;

someone may need to divide humans into "great philosophers and everybody else" or "saints and everybody else". What is important is that the categories of the classification be

1.   complete or exhaustive, i.e. that they add up to one hundred percent of the group classified,
2.   mutually exclusive, with no overlap so that nothing has two niches, and
3.   clear, so that we know where to put things.

A classification which follows the criteria of exhaustion, exclusion and clarity is often called a "partition". So, when we speak in the sequel of the partitioning of data we mean to separate the data in a clear manner into mutually exclusive classes so that each piece of data is in one and only one class.

The first two rules deal with the relation of the several categories among themselves, whereas lack of clarity may affect only individual categories. Not knowing where to put things (lack of clarity) is not the same as having two places to put the same thing (overlapping). Thus, for a library to have the categories of "interesting" and "uninteresting" for its books would not be suitable, since different librarians might not put things in the same place. Of course, in one's personal library it might make perfect sense to have a shelf for one's favorite books.

The secret to constructing a classification as opposed to structuring an existing one with the three structural rules just listed is to find some principle of classification and stick to it at each given stage of the process of division. Usually, the principle of classification can be restated as a question and the possible answers to the question will be the categories. Thus, in our imaginary United Nations library, someone using the principle of language asks "What language is this book in?", and the answers are, of course, automatic. If the answers are not generated automatically, the principle is less than perfect.

The real world--especially that of human artifacts--is a somewhat sloppy place, as our example, when pressed, shows. The existence of bilingual or even polylingual books (e.g. Wittgenstein's *Tractatus*, or a Latin-Greek New Testament) illustrates this. We could, logically speaking, have a separate class for books in more than one language. Or, we could have a convention to the effect that the dominant language is that on the left-hand page, or that which is not the original. The assumption here is that Wittgenstein's books become bilingual only when translated, and that Latin-Greek New Testaments are for people who read Latin better than Greek. In any case, the convention may have a common-sense basis but must be clear in its application.

Changing one's principle during a given stage of a classification will almost always lead to one of the structural defects mentioned. For instance, if we classify books as "fiction, paperback, and in French", we have three principles--content, binding, language. A cheap edition of Simenon's Inspector Maigret stories might be in all three--so that they are not mutually exclusive--while a good quality atlas would not be in any--so they are not exhaustive.

It is conceivable that one divide things up with different criteria. Thus, some banks have machines that automatically give rolls of fifty pennies or fifty dimes when bags of coins are poured in, with no particular reason for putting an individual coin in one roll rather than another, except that it is a penny which arrives at the slot at the right moment. Coin collectors, however, have standards like "mint", "worn", and so forth. People generally need reasons (even bad ones) for what they do. Thus, on a school field trip, it would be possible to load children onto buses on the basis of who wins the dash across the parking lot (like the coin-rolling process). A wise principal, however, would restrain such childish impulses, and have some system for assigning children to buses to be able to keep track of them ("first grade", "second grade", and so forth).

One way of classifying that always works if all else fails is by dichotomy: people are philosophers or not; philosophers are Platonists or not; one knows classical Greek or not; and so on. Note that non-philosophers, non-Platonists, those who fail to know classical Greek--all of these are indefinite concepts. Notice, too, that the indefinite term in a dichotomy is perfectly clear, but a catch-all or miscellaneous term (as in a classification of household objects into furniture, clothes, food, soap, books, and other) is unclear.

Dichotomy often permits us to partition data into multiple classes rather than just two, although we only partition two classes at a time. For example, let us imagine that we want to separate a school class into A students, B students, C students, D students, and F students. Let us say that the rule is dependent on a student's average, M.

M >= 90	is an A;
90 > M >= 80	is a B;
80 > M >= 70	is a C;
70 > M >= 60	is a D; and
M < 60	is an F.

We can use dichotomy as follows:

First, all students are either A students or not-A students. Next, those students who are not A are either B or not-B. Next, those who are not B are either C or not-C. Finally, those who are not-C are either D's or F's. Thus, we have repeatedly used the principle of dichotomy to partition a collection into five mutually exclusive, exhaustive classes.

Brief reflection on the dichotomy method of division shows it to be binary in nature: something either is or is not a member of a set. This method is peculiarly suitable for computing. The following "binary tree" diagram shows the method of dichotomy with respect to the grading sample:

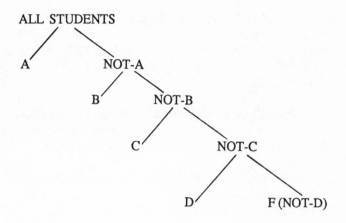

Similarly, we can write a simple program statement to count the number of people with different grades by repeated use of dichotomy:

```
100     IF M >= 90 THEN A=A+1
              ELSE IF M >= 80 THEN B=B+1
              ELSE IF M >= 70 THEN C=C+1
              ELSE IF M >= 60 THEN D=D+1
              ELSE F=F+1
```

where M is the average, and A, B, C, D, F are counters which tell how many are in each category.

The above is an example of an IF...THEN... with the ELSE... option. Recall that IF...THEN... statements have the form:

```
500     IF (expression) THEN (execution)
```

where the (execution) happens when (expression) is TRUE. The ELSE option permits (execution) when the (expression) is FALSE. Thus:

```
500     IF (expression TRUE) THEN (execution 1)
        (expression FALSE) ELSE (execution 2).
```

Statement 100 above demonstrates a nesting of IF...THEN...ELSE... statements.

Notice in the second line, ELSE IF M >= 80... We need not check to see that M<90, since all M>=90 have been eliminated in the first statement.

Exercises: Find three rule violations and say briefly what specifically is wrong with the following. (There may be more than one violation.)

Sample:

Books = philosophic, historical, literary, novels, hardcover.
   a.   Categories are not mutually exclusive. (Literature includes novels, and hardcover includes some or all the other categories.)
   b.   Classification is not exhaustive. There is, for example, no place for a paperback cookbook.
   c.   There is a change of principle at hardcover from content to binding.

1.   Dogs are long-haired and short-haired. Long-haired and short-haired are each for work, hunting or show.
2.   Wines are domestic or imported. Domestic wines are cheap, expensive, or in plastic jugs. Imported wines are red, white, rosé, and sparkling.
3.   Food is in the meat group, the fruit and vegetable group, the cereal group, or belongs to what is served in fast food places.
4.   The French, Germans and artists all ate too quickly.
5.   The house was full of dogs, Dachshunds, cats and panthers.
6.   He ate cookies and quickly.

2.2   Propositions

The categorematic terms of a proposition are its subject and predicate. In "All politicians are friendly", "politicians" is subject

(S, for short) and "friendly" is predicate (P, for short). What connects them is called "copula" (C, for short) in logic. The forms of the verb "to be" are the most common of these last. However, any verb can and should be standardized into copulative form for use in logic. The most common transformation is to change the verb to the present participle form and append the form of "to be". Thus, "barks" becomes "is barking", "ate" can be "was eating", and so on.

In our example, "All" is a quantifier, and "is" is the copula. Other quantifiers are "some", "none", "some are not". Here we find it convenient to use all four quantifiers rather than just "all" and "some", as discussed in Chapter I, above. Any differences between the two notations will be noted as we go along.

A proposition that refers--affirmatively or negatively--to the whole extension of its subject is universal. A proposition that refers to only some of the extension of its subject is particular. The term "universal" is not taken here in some sort of cosmic sense, but only in reference to each term. Thus, although only some dogs are Dobermans, "All Dobermans are vicious" is a universal proposition, referring to the universal set of Dobermans, just like "All dogs have fleas" refers to the universal set of dogs. (We should note that in English, as distinct from some other languages, when a quantifier is not specified, the universal quantifier is usually presupposed. "Water flows downhill" = "All water flows downhill").

We thus have four types of proposition:

Universal affirmative	or "A"	e.g. All roses are red
Particular affirmative	or "I"	e.g. Some roses are red
Universal negative	or "E"	e.g. No rose is red
Particular negative	or "O"	e.g. Some roses are not red

["A" and "I" are from *affirmo*, the Latin word for "assert", while the "E" and "O" are from the Latin word for "deny"--*nego*]

The propositions may be represented graphically by the so-called Venn diagrams:

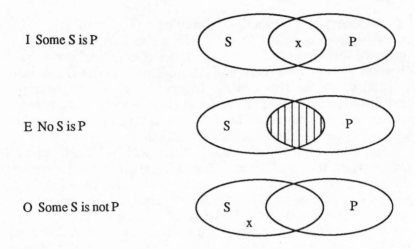

I Some S is P

E No S is P

O Some S is not P

The shaded areas indicate nullity, non-existence. The "x" indicates existence.

Different logical analyses have been made of the A proposition and these can be easily reflected graphically. One must understand that different points of view lead to such variation. For instance, it is possible to consider faculty tenure as a property right (a kind of investment), or as a guarantee of free speech. To the extent that two interpretations are simply different they are compatible and may both be true; in the area where they overlap they will either say the same or be in conflict; and, in the latter case, both cannot, of course, be true.

Aristotle and the medievals understood that to affirm "All roses are red" carries with it the supposition that there are roses. Notice that this is a logical supposition about existence and meaningfulness, not a metaphysical assertion. So, "All snarks are boojums" is meaningful in the imaginary world of Lewis Carroll, as is "All unicorns have a horn" in an imaginary world of unicorns. (In such an imaginary world, unicorns exist, so that we have no problem with the null set, as we did in Chapter I).

A convenient modern analysis of A sentences resolves them into two statements, linked conditionally: "All S is P" becomes "If x is S, then x is P"; "If x is a rose, then x is red". Now, this interpretation permits us to call upon the wealth of structures of the propositional and term calculus of symbolic logic. However, a problem arises with the conditional format in one somewhat counter-intuitive limit case--that of the null class. In truth-table logic, if we are sure that our antecedent "x is S" is not true (which, of course, happens when there is no S), then the larger conditional statement is always true. We can see the difference in the the Venn diagrams for the proposition "All roses are red":

TRADITIONAL

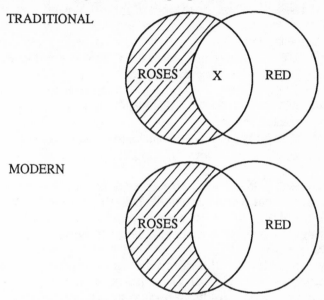

MODERN

In our symbolism, the modern or conditional analysis has "roses" shaded outside of "red", but does not include an "x" in "red"; for there may, at the limit, be no roses at all. Note well that the traditional or "Aristotelian" system has no null set, so "All roses are red" means that there exist roses, while in the modern system the set of roses may be null, i.e. no roses may exist.

The logical form of some sentences in English is not always explicit. In the first place, our standard form is an attribution of a predicate to a subject. Thus, we might rewrite "All dogs chase cats" as "All dogs are creatures (or things) that chase cats" or just "All dogs are cat-chasers".

*   *   *

Before establishing an automatic way of determining the quantifier of any proposition, we must provide for the INPUT of a proposition. In all of our work, we will ask for entire sentences to be INPUT, since propositions must be sentences. Our most trivial and immediate check on sentence input is a period at the end of the INPUT. Thus, our standard inputting routine will be:

```
999     !...INPUT Routine...
100     INPUT "Please type in a sentence, ending with a period, and fol-
          lowed by a carriage return ";W$
1010    L=LEN(W$)\F=0                          ! F is bad-sentence flag
1020    IF RIGHT$(W$,L)<>"." THEN PRINT "You must input a
          sentence"\F=1                        ! = bad sentence
1030    RETURN ! W$ is now the input sentence, and L is its
                          length
```

Of course, just because an expression ends with a period does not mean that it is a sentence! But, these are syntactic and semantic problems, to which we will return in Chapter III. Note that we might also check for a capital letter at the beginning of the sentence.

Now, in order for the computer to determine whether a proposition is an A, E, I or O proposition, it is necessary to check the first word of the sentence and to compare it with "ALL", "NO", or "SOME". Our only problem occurs when we encounter "SOME", in which case we must determine whether it marks an I or an O proposition. This is done by searching for an "is not" or "are not" in a subroutine.

We find the first word of the sentence by finding the first space. Here is a routine to determine the quantification Q$ of a sentence W$ as follows:

```
2000    !...QUANTIFIER Routine...    ! Removes & interprets quantifiers
2110    LET S1=INSTR(1,W$," ")               ! S1 is the first space
2120    LET W1$=LEFT$(W$,S1-1)               ! first word of sentence
2130    IF W1$="ALL" THEN Q$="A"
               ELSE IF W1$="NO" THEN Q$="E"
```

```
              ELSE IF W1$="SOME" THEN GOSUB 3000
              ELSE PRINT "No quantifier found"\GOTO 100
              ! start over if no quantifier found
2140    LET WQ$=RIGHT$(W$,S1+1)  ! this is the sentence, without its
                                                        quantifier
2200    RETURN
3000    !...SUBROUTINE to distinguish I and O...
3110    LET N1=INSTR(1,W$,"IS NOT")
3120    LET N2=INSTR(1,W$,"ARE NOT")
3130    IF N1=0 AND N2=0
              THEN Q$="I"      ! "NOT" not found
              ELSE Q$="O"      ! "NOT" found
3150    RETURN
```

Remember, relative to line 3130 above, that INSTR(1,A$,B$)=O
means B$ has not been found anywhere in A$.

Typical disguised forms of A propositions are "Only the
good die young" and "Whales are mammals". The first is
equivalent to "All who die young are good" (note that it is
necessary to identify the terms correctly to make that reversal).
The second is equivalent to "All whales are mammals". "Blessed
are the poor in spirit" translates into "All the poor in spirit are
blessed".

Disguised I sentences include "Dogs often think they are
human", "A few dogs drink beer", "Most dogs wag their tails
when they are happy", which would be rewritten as "Some dogs
think they are human", "Some dogs drink beer", "Some dogs
wag their tails when they are happy". The explicit copula
separates the S and P terms clearly. Our examples make clear that
a logical term does not have to be represented by one word; very
often, it is represented by a group of words, like "things that wag
their tails when they are happy", which can also be "happy
tail-waggers". A disguised E would be "Whales are never
fish", which is the same as "No whale is a fish". "Sea animals
are seldom mammals" is the O proposition, "Some sea animals
are not mammals". Note that English has a shade of meaning lost
here. "Most sea animals are not mammals" would be a more
precise rendering, but we have only "all" and "some" as logical
quantifiers.

Certain disguised O forms are tricky. "Few logic professors carry pistols" (note not "a few") translates into "Some logic professors do not carry pistols". "All football players do not weigh over 200 lbs." becomes "Some football players do not weigh over 200 lbs." The "All S is not P" construction is a clumsy rendering of "Not all S is P". It occasionally is used as a universal negative and because of this possible ambiguity ought to be avoided as bad English.

Our program for determining quantifiers cannot be used for the disguised forms described above. Certain specific programs can be written for some of the disguised forms, but it is difficult to write a program which will catch all of the disguised forms. Once again, this becomes a semantic problem, to be discussed in Chapter III.

The four types of propositions--A, I, E, and O--are sometimes augmented by affirmative and negative singular propositions. It is possible, however, to treat the singular propositions as universal, since, after all, when there is only one of a kind, it is not possible to talk about part of the group. This simplifies our task and permits a more elegant square of contradiction instead of a hexagon of contradiction.

Exercise 1(Programming):
  1. Alter the previous subroutine so that sentences like "All S are not P" and "All S is not P" are interpreted as E propositions.
  2. Write a subroutine which rewrites "All S is not P" as "No S is P".

Exercise 2: Rewrite the following as a standard A, I, E, or O sentence and provide the appropriate Venn diagram.
Sample:
            Porpoises are mammals.
            All porpoises are mammals.
  1. Only boors smoke cigars during the symphony.
  2. Happy is the man in love.
  3. A few professors are pedantic.
  4. All students are not lazy.

5. Few children pass up pizza.
6. Mothers never abandon their children.
7. Reptiles are sometimes water-dwellers.
8. Cows are never carnivores.
9. Only logicians understand the contrapositive.
10. Professors are often funny.

2.3   Reasoning

2.31   Immediate Inference

Certain precise relationships exist between propositions dealing with the same terms. For instance, "All dogs have fleas" is contradicted by "Some dogs do not have fleas"--one must be true and one false. The same relationship holds between E and I propositions: "No hen has teeth" and "Some hens have teeth" are *contradictories*--if we know that one is true, the other must be false; and, if we are told that one is false, the other must be true.

The relation between universal and particular propositions of the same sign is called *subalternation* (or subimplication). The A proposition ("All cats like tuna") implies the I proposition ("Some cats like tuna"), as long as we are not discussing the null class. The relationship is not symmetrical. If the A proposition is true, so is the I proposition; whereas, if the A proposition is false, the I is undetermined: knowing that it is not true that "All cats wear bells" does not settle anything about "Some cats wear bells". Similarly, if the I is true, the A remains undetermined; knowing "Some Englishmen have bulldogs" does not help us with "All Englishmen have bulldogs". But, if the I proposition is false, as in "Some lobsters are white", the A proposition, "All lobsters are white" must also be false. The same relations hold between universal and particular negatives (E and O).

Contrary opposition holds between complete opposites--e.g. between A and E propositions. "All reindeer have antlers" and "No reindeer have antlers" are contraries.

If either is true the other must be false, but they might both be false, so that knowing one to be false does not help us to determine the truth value of the other; it remains undetermined;

that is, some reindeer may have antlers and some may not.

The I and O propositions are called *subcontraries*; e.g. "Some dentists are Communists" and "Some dentists are not Communists". Obviously, these may both be true. They may not, however, both be false. If it is false that "Some dentists are Communists", that is because "No dentist is a Communist" is true; then, a fortiori, "Some dentists are not Communists" must also be true. Hence, the reverse of what occurs for contraries happens here: if we know that an I proposition is true, the corresponding O is undetermined; but, if the I is false, the O is true.

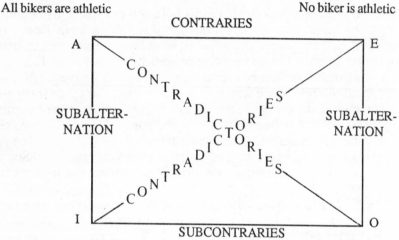

Conversion is an operation which reverses the order of subject and predicate. Three cases can be considered. E and I propositions convert simply to E and I propositions in the reverse order, respectively. Thus "No cat likes to swim" converts to "Nothing that likes to swim is a cat" or "No swimmer is a cat" and "Some dogs weigh over 100 lbs." converts to "Some things that weigh over 100 lbs. are dogs" or "Some 100-pounders are dogs". One can convert back to the original propositions since an E and I proposition and their respective converses are strictly equivalent.

By contrast, A propositions do not convert simply, but become I propositions. "All goats have hooves" does not imply that "All hooved animals are goats", for there are other hooved animals. One of the peculiarities of A propositions is that their predicates are not taken in their full universal extension but are particular (also called "undistributed"). So, the converse of "All goats have hooves" is "Some hooved animals are goats", since "goats" is taken universally but "beasts with hooves" is particular.

O propositions have no converse. Intuitively, one can see that "Some Roman Catholics are not Cardinals" would not imply that "Some Cardinals are not Roman Catholics". "Some lions do not live in zoos" and "Some animals in zoos are not lions" happen both to be true but have nothing to do with one another. In addition, we see that the converse of the true proposition "Some dogs are not Dobermans" would be the false proposition "Some Dobermans are not dogs" (of course, we are not talking about Mr. and Mrs. Doberman!).

In order for us to manipulate sentences, we must be able to find the subject and predicate of a given sentence. From the previous routines, we have sentence W$ with quantifier Q$, while WQ$ is the same sentence without the quantifier. (There should always be a space between the quantifier and the subject). For our present purposes, all W$ must have "is" or "are" as the copula. Thus, all of our sentences will be of the form

W$=Q$+S$+C$+P$,

where WQ$ is the sentence and

Q$= the quantifier
S$= the subject
C$= the copula ("is", "is not", "are", "are not")
P$= the predicate

So, we must write a subroutine to find the copula C$. Once this is done, everything to the left of C$ (not including the quantifier Q$) is the subject S$, while everything to the right of C$ is the predicate P$.

```
4000    !...SP-ANALYZER...
4010    ! SUBROUTINE TO FIND SUBJECT AND PREDICATE
4100    ! To find whether the copula is "is" or "are",
4110    ! we initialize everything for "is" and change them for "are"
4200    LET C$="IS"                        ! assume copula to be "is"
4210    LET CS=4                           ! COPULA-SPACES = 4
                                                   ...is...
4220    LET IC,AC=0                        ! set location of "is" or
                                                   "are" to 0
4300    LET IC=INSTR(1,WQ$," IS ")
4310    IF IC=0 THEN LET AC=INSTR(1,WQ$," ARE ")\LET
            CS=CS+1\C$="ARE"               ! add space for extra letter
                                             & put "are" for "is"
4320    IF AC+IC=0 THEN PRINT "No copula found"\RETURN
4400                                       ! separate the subject
                                                   (below)
4410    S$=LEFT$(WQ$,IC+AC)                ! "IC+AC" is the maxi-
                                             mum of IC or AC since
                                             one of them is 0
4500                                       ! now separate the predicate
4510    IF Q$="O" THEN LET CS=CS+3\C$=C$+"NOT " ! adds "NOT"
                                                   to copula
4520    LET P$=RIGHT$(WQ$,IC+AC+CS)  ! predicate
4900    RETURN
```

Once we have found the quantifier Q$, the subject S$, the copula C$, and the predicate P$, a simple subroutine can be written for doing the conversion of propositions, as follows:

```
5000    ! ...CONVERSE Routine...
5100    IF Q$="A" OR Q$="I"                ! the converse of "ALL"
            THEN CON$="SOME "+P$+C$+S$ ! or "SOME"
                                                   is "SOME"
                ELSE IF Q$="E"            ! "NO" is its own
                THEN CON$="NO"+P$+C$+S$        ! converse!
                    ELSE CON$=" "    ! since O has no converse
5200    RETURN
```

given proposition, using subroutines we have already developed.

```
10      ! ...Program to find and print CONVERSE...
20      !
100     GOSUB 1000                    ! get the sentence
110     IF F=1 THEN 100               ! bad sentence: start over
200     GOSUB 2000                    ! determine the quantifier
210     GOSUB 4000                    ! find the subject and
                                                  predicate
220     GOSUB 5000                    ! find the converse
230     IF CON$=" " THEN PRINT "The proposition is not convertible"
                ELSE PRINT CON$;" is the converse of ";W$
300     STOP
```

Obversion involves the principle that two negations cancel each other out. The propositions change quality (not quantity, however) and definite predicates become indefinite. It is always possible to obvert. A proposition is strictly equivalent to its obverse. They are true or false together. Hence we have:

All roses are red	=	No rose is other than red.
Some roses are red	=	Some roses do not fail to be red.
No rose is red	=	All roses are non-red.
Some roses are not red	=	Some roses are other than red.

"Other than", "non-", "fails to be" are simply different ways of articulating indefinite concepts.

A subroutine for obversion is quite simple once we choose which of the articulations we want to use. Without loss of generality, we choose "non-" for simplicity. As usual, Q$=the quantifier, S$=the subject, C$=the copula, P$=the predicate, and are to be found in the SP-ANALYZER.

```
6000    ! ...OBVERSE Routine...
6100    IF Q$="I" OR Q$="O"
                THEN OB$="SOME"+S$+C$+"NON-"+P$
                ELSE IF Q$="A"
                        THEN OB$="NO "+S$+C$+"NON-"+P$
                ELSE IF Q$="E"
```

    ELSE OB$=" "   ! which can happen if there
                                        is no sentence at all!

6200   RETURN

Exercise: Write a program to print out the obverse of any statement.

The *contrapositive* is shorthand for the obverted converted obverse. There are only three contrapositives because when one obverts an I proposition, it becomes an O, and the second step, conversion, cannot be done. The A and O propositions are equivalent to their contrapositives--that is, the A and O propositions and their respective contrapositives are either both true or both false. The E proposition has an O contrapositive which it implies; the truth conditions are the same as for subalternation. Here, then, are the three contrapositives:

All violets are blue   = All non-blue things are other than
                                     violets
Some dogs do not bark   = Some things that fail to bark do not
                                       fail to be dogs
No pig has wings (implies)   = Some thing that lacks wings is not
                                        other than a pig

Here are the steps by which the contrapositive of an A proposition is derived:

1.   All logicians have a wry sense of humor.
2.   No logician fails to have a wry sense of humor (obverse).
3.   No one who fails to have a wry sense of humor is a logician (converse).
4.   All who fail to have a wry sense of humor are non-logicians (obverse).

That is, the contrapositive is simply the obverted converted obverse.

Since the contrapositive can be defined through the converse and obverse, we have almost all of the necessary program code developed in the previous sections: for all we need do is INPUT a sentence, find its *obverse*, find the *converse* of the obverse, and find the *obverse* of the converted obverse. The following program uses all the programs we have developed in this chapter to find the contrapositive (note that the line numbers match those in already developed programs).

```
10       ! ...Program CONTRAPOSITIVE
20       ! Glenn Satty         July 14, 1987
30       ! The purpose of this program is to find the
40       ! contrapositive of a given proposition.
50       ! This program uses the subroutines INPUT,
60       ! Q-ANALYZER, SP-ANALYZER, OBVERSE and
70       ! CONVERSE, as previously developed
100      GOSUB 1000                          ! input the sentence
110      IF F=1 THEN 100                     ! bad sentence, restart
120      LET D$=W$                           ! store the original
                                                 proposition
200      GOSUB 2000                          ! determine quantifier
210      GOSUB 4000                          ! find subj. and pred.
220      GOSUB 6000                          ! find obverse
230      IF OB$="" THEN PRINT "The proposition is not obvertible"
             \GOTO 800
240      LET W$=OB$                          ! now, W$=obverse of
                                                 original
300      GOSUB 2000                          ! new quantifier
310      GOSUB 4000                          ! new subject and predicate
320      GOSUB 5000                          ! find converse of obverse
330      IF CON$=" " THEN PRINT "The obverse is not convertible"
             \ GOTO 800
340      LET W$=CON$                         ! now, W$ is converse of
                                                 obverse
400      GOSUB 2000
410      GOSUB 4000
420      GOSUB 6000
430      IF OB$=" " THEN PRINT "The proposition has no contrapositive"
             \ GOTO 800
440      LET W$=OB$                          ! W$ is now the
                                                 contrapositive
```

```
500      PRINT W$
510      PRINT " is the contrapositive of "
520      PRINT D$                               ! the stored original
800      PRINT
810      PRINT "Would you like to do another sentence?"
820      INPUT "If so, please type Y ";A$
830      IF A$='Y' OR A$='y' THEN 100
900      STOP
1000     ! fill in the rest of the lines
2000     ! from the corresponding parts
3000     ! of previous programs and
4000     ! routines
5000
6000
9999     END
```

Exercise: Many times, the contrapositive may contain "non-non-" or "not non-". Write a routine to eliminate any double negations like these.

Here are the steps by which the contrapositive of an O proposition is derived:

1. Some logicians cannot afford fine wines.
2. Some logicians are other than those who can afford fine wines.
3. Some who are other than those who can afford fine wines are logicians.
4. Some who are other than those who can afford fine wines are not other than logicians.

For practice, the reader might work out these relations for some simple propositions for the O and E cases.

The converse of the contrapositive of an A sentence is called its "inverse". Since the contrapositive of both E and O propositions are O in type, they cannot be converted--so E and O have no inverse.

As a memory aid, here in symbols are the relationships we have been discussing:

	A proposition	**E proposition**
	All S is P	_No S is P_
contrary	No S is P	All S is P
contradictory	Some S is not P	Some S is P
converse	Some P is S	No P is S
obverse	No S is non-P	All S is non-P
contrapositive	All non-P is non-S	Some non-P is not non-S
inverse	Some non-S is non-P	(no inverse for E)

	I proposition	**O proposition**
	Some S is P	_Some S is not P_
subcontrary	Some S is not P	Some S is P
contradictory	No S is P	All S is P
converse	Some P is S	(no converse for O)
obverse	Some S is not non-P	Some S is non-P
contrapositive	(no contrapos. for I)	Some non-P is not non-S

As a drill, the reader can choose terms as values for S and P, and translate the above symbols back into English. So if S=mule, and P=kicks, "All S is P" would be "All mules kick" and "Some P is S" would be "Some things that kick are mules". The reader can choose whatever values for S and P that strike his or her whimsy or aid his or her intuition.

Exercises:
1. Write a program to determine the contrary, contradictory and proper "altern" of any given proposition (with copula "is" or "are"). Use any subroutine you may need from the text.
2. In each of the following:
   a. Put in standard form
   b. Symbolize
   c. State the relation, if any
   d. If I is true, what about II?
   e. If I is false, what about II?

Sample:
0. I. A few cats eat mice.
   II. Cats never eat mice.
      a. Some cats are mice-eaters.
         No cat eats mice.
      b. Some S is P
         No S is P
      c. contradiction
      d. false
      e. true

1. I. Only bad dogs chase cats.
   II. All bad dogs chase cats.
2. I. Cats never swim.
   II. Cats sometimes do not swim.
3. I. All cats eat tunafish.
   II. No cats fail to eat tunafish.
4. I. All cats go crazy with catnip.
   II. A few cats go crazy with catnip.
5. I. Cats wear bells.
   II. Some cats do not wear bells.
6. I. Few cats are striped.
   II. A few cats are striped.
7. I. All cats are not selfish.
   II. All cats are selfish.
8. I. Tabby cats are never cute.
   II. Tabby cats are always cute.
9. I. Only women like cats.
   II. Some women like cats.
10. I. Only eccentrics like Manx cats.
    II. All who are other than eccentrics fail to like Manx cats.
11. I. All Persian cats fail to purr.
    II. No Persian cat purrs.
12. I. No Abyssinian cat has long hair.
    II. Some things that are other than long-haired are not other than Abyssinian cats.

In the following let S=philosophers and P=crazy, and translate the formulae back into English.

Sample:

> All S is P = All philosophers are crazy.
> No non-P is non-S = No people who fail to be crazy are other than philosophers.

1. No S is P.
2. Some S is P.
3. Some S is not P.
4. Some P is S.
5. Some P is not S.
6. All S is non-P.
7. No S is non-P.
8. All non-S is non-P.
9. All non-P is non-S.
10. No P is S.

2.311 Modal Logic

It is helpful to review some of the basic notions of modal propositions. By "modal proposition" is meant a statement which is not simply affirmed as true or false, but as true or false with a peculiar degree of certainty. Thus a proposition may be

> necessary or not
> possible or not

A necessary proposition is also possible. An impossible proposition is also non-necessary or, to use the proper term, "contingent". Here are some examples in ordinary language:

> What goes up must come down. -- necessity
> The Boston Red Sox may win the 1990 pennant. -- possibility
> It is not certain that I will live in retirement. -- contingency
> There can be no such thing as a square circle. -- impossibility

These four notions can be arranged in the traditional square of contradiction:

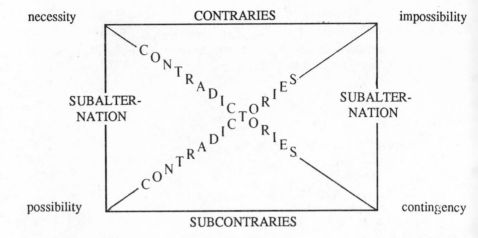

necessity   CONTRARIES   impossibility

SUBALTER-   SUBALTER-
NATION   NATION

possibility   contingency
                            SUBCONTRARIES

Matters get more complicated when we combine modal operations and quantifiers. Our four basic types--A, I, E, O--combined with the four modes, give us sixteen types of quantified modal propositions, some of which, however, are equivalent, and some not clearly opposed. "It is necessary that all men die" (i.e. "All men must die") is contradicted by "It is possible that some men not die", or its equivalent, "It is not necessary that some men die". "It is possible for all philosophers to be confusing" is contradicted by "It is impossible for some philosopher to be confusing". "It is possible for some philosopher not to be confusing" is quite compatible with the first statement.

The mechanics of this now--not a square but a decasexagon of opposition--are bad enough at the intuitive level; but the subject of modal logic becomes completely murky when with the best intentions in the world it is applied to modern symbolic logic. It is tempting to say that logical theorems (e.g. p V ~p) are necessarily true, whereas p•~p is impossible and (p) is contingent. Indeed, the American logician C.I. Lewis invented a special operator for strict implication--or, as some might say, real implication--to capture what we mean when we say "That a number in the series of integers is even implies that its successor is odd". The definition of implication in truth-tables (oddly called "material" implication) allows for the somewhat bizarre extension of the relation of implication to a number of situations, including any two propositions, as in: "That Ulan Bator is the capital of Outer

Mongolia implies that I.M. Bochenski wrote *A History of Formal Logic*". The kindest thing that one can say of this use of "imply" is that it is counter-intuitive, so that Lewis seems to be onto something.

Unhappily, attempting to develop a formal system of modal logic has usually led to paralleling the mechanisms for any formal logic and thus defeating the purpose of the whole endeavor.

Exercise:
Sample: Give the contradictory of "It is possible for Ronald Reagan to run for President in 1988".
Answer: "It is impossible for Ronald Reagan to run for President in 1988."

1. Give the contrary of "It is necessary for Reagan and Gorbachev to reach an arms agreement."
2. Give the contrary of "It is possible for Reagan and Gorbachev to reach an arms agreement."
3. Contradict "It is impossible for any logician to commit a fallacy ever."

2.32   The Assertoric Syllogism

Aristotle (384-322 B.C.) in ancient Greece worked out a complete explanation of the syllogism, a kind of simple reasoning, where a conclusion is drawn from two propositions. To understand the theory of the syllogism, it is necessary to realize that ordinary language is rather free in its order, whereas logic has a rigid standard format for purposes of clarity in syllogisms, much as it has a rigid standard format for the four types of propositions.
The standard order for the syllogism is:

1. the premiss with the predicate of the conclusion (known as the *major* term) is the *major* of the syllogism;
2. the premiss with the subject of the conclusion (known as the *minor* term) is the *minor* of the syllogism;
3. the conclusion, where the subject is S and predicate is P.

Obviously, we must identify the conclusion first. Words like "therefore", "thus", "accordingly", "so", "it follows", "as a result", "consequently", "hence", "then" indicate the conclusion. Words like "since", "because", "for", "in as much as" indicate the premises. For example, "Since all cats are felines, all cats are carnivores, for all felines are carnivores" becomes:

> All felines are carnivores
> <u>All cats are felines</u>
> So, All cats are carnivores

where S=cats and P=carnivores. Or, again, "No little boy likes spinach because all who like spinach are responsible eaters and no little boy is a responsible eater" becomes:

> All who like spinach are responsible eaters
> <u>No little boy is a responsible eater</u>
> So, No little boy likes spinach

where S=little boy and P=likes spinach.

It should be noted that the syllogism has three terms. Besides the S and P terms (minor and major terms), one term appears in both premises but not in the conclusion. It is called the middle term. In the first example above the middle term (or M) is "felines" and in the second "responsible eater". Our search here is to determine whether a syllogism is formally valid. That is, is the conclusion necessarily true if the two premises are true?

We now have two rules to check on whether an argument constitutes a valid syllogism: the middle term should not appear in the conclusion, and there should be exactly three terms. Two terms do not give a new proposition as conclusion, nor would the appearance of the middle term in the conclusion; four terms give no linkage or basis for a conclusion, because "if a=b and b=c, then a=c"; i.e. unless there is a "b" to link "a" and "c" one cannot speak of "syllogism".

Thus, a syllogism cannot be valid unless:

1. It has only three terms;
2. The middle term does not appear in the conclusion but appears once in each of the premises.

There are four basic arrangements of terms in the premisses--called the four *figures*. We can represent these as follows:

I	M P	II	P M	III	M P	IV	P M
	S M		S M		M S		M S
	S P		S P		S P		S P

Since either premiss can be any one of the four kinds of proposition, we have a total of sixteen conceivable modes of syllogism for each of the four figures. Of the 64 conceivable modes, only nineteen are valid. If we were to consider the conclusion as being one of the four kinds of proposition as well, we end up with 256 (see Appendix A) conceivable modes, but still only 19 valid ones. To work out which ones are valid, it is necessary to know only a few rules.

In what follows, the twelve significant terms of the syllogism (three each of quantifiers, subject-terms, copulae, and predicate-terms) will be symbolized as follows:

$$A\$ \quad B\$ \quad C\$ \quad D\$$$
$$E\$ \quad F\$ \quad G\$ \quad H\$$$
$$I\$ \quad J\$ \quad K\$ \quad L\$$$

Thus, A$, E$, and I$ are the three quantifiers; C$, G$, and K$ are the copulae, and so on.)

The easiest rules to apply are those that prohibit both premisses from being negative or both from being particular. Thus, from the premisses

No frog is a fish
No dog is a frog

nothing follows. The Venn diagrams will show us the flaws in the syllogism with two negative premisses.

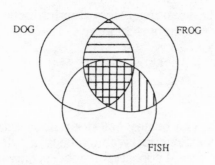

DOG · FROG · FISH

This mode of syllogism is invalid because of the many possible implications. Nor does anything follow from

> Some deer are dwarf animals
> Some animals in this zoo are deer

Evidently, the animals in this zoo may or may not be dwarf.

Between them, the rules prohibiting two negative premisses or two particular premisses eliminate some 112 conceivable modes of syllogism. Note that "OO" syllogisms are prohibited by both rules.

The exclusion of the double negative can be represented as:

IF A$="NO" AND E$="NO" THEN GOTO 20000
IF A$="NO" AND E$="SOME" AND G$="<>" THEN GOTO 20000
IF E$="NO" AND A$="SOME" AND C$="<>" THEN GOTO 20000

where 20000 would be the line that says "No conclusion follows from two negative premisses". Note that we have excluded not only E + E but also E + O and O + E.

Strictly speaking, the last example about dwarf deer has another defect. The middle term is taken particularly (undistributed) in both premisses. Remember that a term is particular or undistributed either because it is the subject of a particular proposition or the predicate of an affirmative proposition. Hence, from

> All birds live in trees
> All squirrels live in trees

we have no business concluding that all squirrels or any squirrels are birds. Only part of the tree dwellers are birds and part are squirrels, and the premisses give us no indication that they are the same part.

Venn diagrams show the problem of the particular (undistributed) middle as follows:

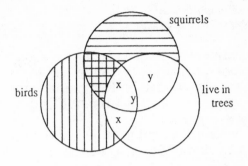

The premisses do not permit an unequivocal conclusion, and logic is interested only in unequivocal conclusions. That is, when undistributed, the middle term can be divided into two parts--one belonging to the S term and the other to the P term, and we really have four terms instead of three (although two are identical in name but not in meaning). So, an undistributed middle does not necessarily connect the premisses the way a middle term must for a syllogism to be valid.

By contrast, terms that are the predicate of negative propositions are universal (distributed). We cannot say "Some swan is not white" unless we exclude that swan or swans from the whole extension of the term "white". Of course, subjects of universal propositions are also universal. This principle must be recalled when we make sure that the minor and major terms are not used illicitly; i.e. that we do not have information about only "some S" or "some P" in the premiss, but presume to know about all S or P in the conclusion. Obviously, we would not want to say:

All Hegelians believe in dialectical logic
Some philosophers are Hegelians
So, All philosophers believe in dialectical logic

However, we would (formally speaking) commit the same error in

>All Communists admire Lenin
>All Communists are Marxists
So, All Marxists admire Lenin

We have no business drawing a conclusion about "all Marxists" since our minor premiss only entitles us to say that "some Marxists are Communists". Remember, the conclusion (predicate), of a universal affirmative is always taken as particular, not as universal.

In fact, the same general principle is violated again in this example in the illicit use of the major term:

>All wine is made from grapes
>No drink for children is wine
So, No drink for children is made from grapes

The major premiss allows that there may be other things besides wine made from grapes, but the term "made from grapes" in the conclusion must be taken in all its extension, or the negation makes no sense at all. It must be affirmed that this would be true even if the conclusion were a particular negative.

Venn diagrams illustrate the equivocal nature of an inference based on an illicit major:

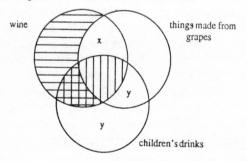

The sector common to "children's drink" and "things made from grapes" is blank, not shaded. We just do not know about it, as the authentic diagram with three circles shows.

A syllogism with a negative premiss must have a negative conclusion and a syllogism with a particular premiss must have a particular conclusion; a syllogism with both must have a negative particular conclusion. The following three syllogisms are valid:

> No surgeon is poor
> All philosophers are poor
> It follows that no philosopher is a surgeon

> All bureaucrats take coffee breaks
> Some Republicans are bureaucrats
> Therefore, some Republicans take coffee breaks

> No geranium is edible
> Some plants in my house are geraniums
> So, Some plants in my house are not edible

Evidently, we could not conclude that any philosophers are surgeons, or that all Republicans take coffee breaks, or that some plants in my house are edible, or anything about all my plants. The conclusion cannot be stronger than the weaker premiss; which is to say we cannot conclude "All" or "No" if we only start with "Some", or an affirmative if we start with a negative.

On the other hand, if we have two affirmative premisses, we must have an affirmative conclusion. In the following fourth-figure syllogism

> All Trotskyites are infantile leftist wreckers
> All infantile leftist wreckers should be purged
> So, Some who should be purged are Trotskyites

the possibility is left open that there are others (perhaps Mensheviking idealists or capitalist roaders) who should also be purged. The premisses could be rearranged to get a more simple first-figure syllogism, the conclusion of which would involve all Trotskyites, but no negative conclusion can be drawn.

Notice that some modern logicians either reject this figure or call it only "conditionally valid". Traditional logic understood that "All Trotskyites are infantile leftist wreckers" supposes the existence of Trotskyites. By contrast, the nineteenth-century translation of the proposition as "If there are Trotskyites, then they are infantile leftist wreckers" left the proposition true even if there are no Trotskyites. This is certainly not the normal point of the affirmation, and it might even be argued that it is counter-intuitive and hence a deformation of ordinary language.

The rule that a negative cannot follow from two affirmatives is of no interest in the second figure, where there must be one negative to avoid the undistributed middle (which would occur if the middle term were predicate of an affirmative proposition in both instances). In the third or first figure a negative conclusion from two affirmative premisses would entail an illicit major since the predicate of the negative conclusion (always universal) initially appears as the predicate of an affirmative premiss (always particular). Strictly speaking, since the rule can be proven from another rule it is superfluous (although helpful) for first and third figure syllogisms.

The fourth figure is always less obvious. Consider

> All logicians are witty
> <u>All witty people are amusing</u>
> So, Some amusing people are not logicians

This conclusion is, in fact, true and also true will be *de facto* most inferences drawn along this format. It is not, however, validly inferred from the premisses. The conclusion does not violate the obvious rules: i.e. the major term is given universally in the premiss and the middle term is distributed once. But, the syllogism would require information about those amusing people who are not witty and hence not logicians. There may (or may not) be such people, but the premisses do not describe them.

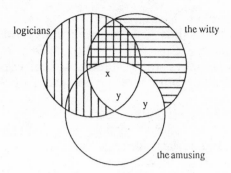

The alleged conclusion requires this diagram, where "y" would be those amusing people who are not logicians. We are given no such group.

We can now formulate a checklist for validity. Five points relate to the premisses as propositions and five relate to the terms. If any of these rules is violated, the syllogism is invalid. Although in some cases, more than one rule can be violated, with practice, the student will realize that some of the rules about propositions contemplate situations that are mutually incompatible (e.g. two affirmative premisses versus two or even one negative premiss), so that, if one applies, the other is irrelevant. These questions must be asked:

1. If both premisses are affirmative, is the conclusion affirmative?
2. Are there two negative premisses?
3. Are there two particular premisses?
4. If we have one negative premiss, is the conclusion negative?
5. If we have one particular premiss, is the conclusion particular?
6. Are there exactly three terms in the premisses and no new term appears in the conclusion?
7. Does the middle term appear in the conclusion?
8. Is the middle term universal (distributed) at least once?
9. Is the major term particular in the premiss and universal in the conclusion ?
10. Is the minor term particular in the premiss and universal in the conclusion ?

Exercises:
   a. Put the following syllogisms in standard form and order
   b. Symbolize

c. Say whether the item is valid or not
d. State the figure of valids or the reason for invalidity.

Sample: Only philosophers read Hegel, so only philosophers understand Marx since only those who read Hegel understand Marx.
a. All who read Hegel are philosophers
    <u>All who understand Marx are people who read Hegel</u>
    All who understand Marx are philosophers
b. All M are P
    <u>All S are M</u>
    So all S are P
c. Valid
d. First figure
1. Since all philosophers drink hemlock and all who drink hemlock irritate Athenians, it follows that only philosophers irritate Athenians.
2. Philosophers always have beards because people who think a lot push the hair out of their heads and philosophers think a lot.
3. No existentialist wears college colors, for they all wear black sweaters, and people who wear black sweaters are serious.
4. Because a few theologians are pantheists, it follows that some pantheists are rigorous thinkers, for some theologians are rigorous thinkers.
5. No Aristotelian is an idealist, because all Platonists are idealists and no Aristotelian is a Platonist.
6. All Prussians are Kantians, because all Kantians are devoted to an ethic of duty and only Prussians are devoted to an ethic of duty.
7. Since some Englishmen are Hegelian and no empiricist is a Hegelian, some Englishmen are not empiricists.
8. Some pragmatists are sceptics, since all sceptics are relativists and all pragmatists are relativists.
9. All Leninists are Marxists and all Stalinists are Leninists, so accordingly all Marxists are Stalinists.
10. No angel dances on the head of a pin, so no angel is a virus, since all that dances on the head of a pin is a virus.

2.321 The Enthymeme

In ordinary conversation and even in writing, a good deal of syllogistic reasoning goes on but is not fully explicated. We might have

> Cocker spaniels are gentle
> Therefore, they make good pets

Here the major premiss--"gentle animals make good pets"--is left unstated. When either premiss or the conclusion is tacit, the syllogistic reasoning is called enthymeme. Enthymemes obey the same rules as syllogisms. Accordingly,

> All lizards are reptiles
> Therefore, no snake is a reptile

is invalid. This enthymeme assumes the minor premiss that "no snake is a lizard"; but, whatever it assumes, the P term, "reptile", is used illicitly since it is given as the predicate of an affirmative proposition (and thus is undistributed) but ends up as the predicate of a negative proposition (and thus is distributed). So this is a disguised AEE-1 which is invalid.

Sometimes, an enthymeme can be misleading because what is omitted sounds like something true, but what would have to be assumed to make the syllogism correct is not true. For example,

> All Communists interfere with private property
> Therefore, all authoritarians interfere with private property

might sound vaguely plausible because there is some connection between Communists and authoritarians. The trouble is that to make the enthymeme valid, we would need to suppose "All authoritarians are Communists" which is certainly not true, and not just "All Communists are authoritarians", which is at least arguable.

Either major premiss or minor premiss or conclusion may be left unexpressed--usually because they are obvious and their explicit enunciation would be tedious. However, in all enthymemes there exists the danger that the missing portion sound like something that is true so that, until it is made explicit, we do not see that the particular enthymeme is flawed. Something outrageous might be needed to make the syllogism valid; it is the outrageous statement that sounds like something true but will not make the syllogism valid. As long as part of the syllogism--especially the conclusion--is not expressed, the difference may not be detected. Consider "All bats are vertebrates, so all bats are mammals." The missing premiss must have something to do with mammals and vertebrates and, in fact, we know "All mammals are vertebrates". But, let us look more carefully:

> All mammals are vertebrates
> <u>All bats are vertebrates </u>
> So, All bats are mammals

This syllogism suffers from an undistributed middle. To make it logically valid, we need "All vertebrates are mammals" which, of course, is false, although the syllogism AAA-1 is valid. Thus, it is important to reconstruct the missing part of the syllogism.

Let us first look at an enthymeme which lacks a premiss, i.e. where we have one premiss and the conclusion. The procedure is:

1. We identify the conclusion by the logical words indicating either inference or reasons, as we indicated above in relation to the syllogism. (see Section 2.32)
2. The conclusion's predicate is the major term and tells us which premiss is the major premiss.

For example: "Some honor roll students hate spinach because all little boys hate spinach". "Because" introduces a premiss; so, "Some honor roll students hate spinach" is the conclusion. Since the predicate of the conclusion appears in the premiss we have, it is the major premiss and its other term is the middle term. (If both subject and predicate of the conclusion appear in one premiss, there is no syllogism because there is no middle term.) We now have:

> All little boys hate spinach $\qquad\qquad$ All M is P
> [$\qquad\qquad\qquad\qquad\qquad$]
> Some honor roll students hate spinach $\qquad$ Some S is P

What precisely is missing? Obviously, a proposition linking "honor roll students" and "little boys". What else do we know?

A. Since the conclusion is affirmative, we need an affirmative premiss (A or I).
B. As the conclusion is particular, we will probably find a particular (I) premiss.
C. The term "little boys" is already taken universally in the premiss: so, it need not worry us in the other premiss.

So, we could try either "Some honor roll students are little boys" or "Some little boys are honor roll students"--both of which will give us valid syllogisms, in the first and third figures, respectively. In this case, universal propositions would also give us valid syllogisms; but, there is no point in assuming statements any stronger than we need, unless they happen to fit the needs of our application.

The greatest single danger in such enthymemes is the invalid major due to a negative conclusion. Consider:

All penguins lay eggs	All M is P
[ ]	
So, no seal lays eggs	No S is P

What seems to be missing is a statement about seals; that is, the minor premiss that would be negative (and universal) because the conclusion is negative, while the premiss we have is affirmative. So, it seems we are looking for "No seal is a penguin" or "No penguin is a seal"--which, by the rules of conversion we saw above, are equivalent. The problem is, however, that nothing will make this enthymeme valid. "Lay eggs" in the premiss is the predicate of an affirmative proposition; hence it is particular. However, it appears in the conclusion either as predicate of a negative proposition (therefore universal) or the subject of a universal proposition (therefore universal). We have here an invalid major.

Other invalid enthymemes are more obvious:

All my animals are pigs	All M are S
[ ]	
So, All pigs are omnivores	All S are P

"Pigs" is particular in the premiss and universal in the conclusion.

There are times when an enthymeme cannot be made valid--i.e. we get caught between two rules. Take the following seemingly innocent case:

All aspirin are bitter   All P are M
[   ]
So, All my pills are aspirin   All S are P

The minor premiss is missing. It must be universal, affirmative, and involve the terms "my pills"(S) and "bitter"(M). But, if we try "All my pills are bitter",

All aspirin are bitter   All P is M
<u>All my pills are bitter</u>   All S is M
So, All my pills are aspirin   All S is P

we get an undistributed middle. On the other hand, if we try "All bitter things are my pills", which is silly,

All aspirins are bitter   All P are M
<u>All bitter things are my pills</u>   All M are S
So, All my pills are aspirin   All S are P

we get an illicit minor term "pills" which is particular (predicate of an affirmative proposition) in the premiss but universal in the conclusion.

The exercises have made us familiar with valid forms of the syllogism so that we can see that the major here requires that we have either a second or fourth figure syllogism, but that all second-figure syllogisms have negative conclusions, and all fourth-figure affirmative conclusions are particular. There is nothing to be done.

For the other instance of enthymeme--where the conclusion is missing--we do not need to go to the same lengths, but we do have to find out which is the major premiss and which is the minor premiss (it will often not matter which is which). For example, in the first figure:

No cat likes baths   No M is P
<u>All witches' pets are cats</u>   All S is M
So, No witches' pets like baths   No S is P

while, in the fourth figure:

All witches' pets are cats	All P is M
No cat likes baths	No M is S
So, No bath-liker is a witches' pet	No S is P

There are other instances, where there are significant differences according to the order of the propositions. In the first figure:

All philosophers love logic	All M are P
All Aristotelians are philosophers	All S is M
So, All Aristotelians love logic	All S is P

we get a universal conclusion, while in the fourth figure we find:

All Aristotelians are philosophers	All P is M
All philosophers love logic	All M is S
Some logic-lovers are Aristotelians	Some S is P

Of course, even here the conclusions are related by conversion.

Question: Could there ever be a case where the ordering of the premisses would be the difference between validity and invalidity?
Answer: The last two examples give us a clue. The different ordering gave us conclusions which were each other's converse. But, O propositions have no converse. Whereas in the first figure we have

No cat likes to swim	No M are P
Some pets are cats	Some S are M
So, Some pets do not like to swim	Some S are not P

there is no fourth-figure syllogism to parallel this.

| Some pets are cats | Some P is M |
| No cat likes to swim | No M is S |

has no conclusion, for we cannot use "pet", when particular, as the predicate of a negative proposition.

Exercises:
   a. Put into standard form and format.
   b. Complete so as to make valid, if possible.
   c. Symbolize (hint: start symbolizing after step a.)
   d. Explain the incompletables and name the figure of the completeds.
Sample: No man lives forever, so all cartoon characters live forever.
   a. No man is one who lives forever   [...]   So all cartoon characters are ones who live forever.
   b. This is not completable.
   c. No M is P [...] So all S is P
   d. You cannot get an affirmative conclusion when one premiss is negative.
1. Some dogs have fleas since all animals that romp in the woods have fleas.
2. No dog is disloyal, so no dog abandons his master in the snow.
3. All animals that carry brandy are friendly, therefore all St. Bernards carry brandy.
4. Dachshunds hunt in burrows so no dachshund has long legs.
5. Some dogs are nearsighted, because all animals in this clinic are dogs.

2.33 The Hypothetical Syllogism

The syllogisms we have just studied are sometimes called categorical syllogisms to distinguish them from hypothetical syllogisms, which feature conditional statements as their major premise and which depend on the relations of implication that we have already encountered in our discussion of immediate inferences. The basic units or elements of categorical syllogisms are terms. The elements, by contrast, of hypothetical syllogisms are propositions. The inferences we draw depend on the relations represented by syncategorematic connectors: in the first place, "if...then...", but also "...or..."
   Two types of inference can be drawn from a conditional sentence. We can affirm that the condition (or antecedent) is fulfilled and thus affirm the result or consequent. Or, we can deny that the result or consequent obtains and hence be forced also to deny that the antecedent

obtains. Thus, we have the *modus ponendo ponens*:

> If the Federal Reserve Bank increases the discount rate, prime rates increase.
> The Federal Reserve Bank increases the discount rate.
> So, prime rates increase.

and *modus tollendo tollens*:

> If the Federal Reserve Bank increases the discount rate, prime rates go up.
> The prime rate has not gone up.
> So, The Federal Reserve Bank has not increased the prime rate.

Strictly speaking, it is not valid to draw a conclusion on the basis of negating an antecedent or affirming a consequent. In fact, however, both in ordinary conversation and even in the sciences, we do something like this. The mother who warns her child "If you get your feet wet, you will catch cold" may be apt to conclude when the child catches cold (i.e. the consequent is affirmed) that he or she got wet feet (affirm the antecedent). Of course, the child may have sat in a draft or forgot his or her hat.

Scientists behave somewhat like the mother when they confirm their hypotheses, except that they are supposed to predict many different consequences, so that they begin to approximate in their confirmation of predictions, a kind of equivalence or "if and only if" relation. The difference between these relations will be made clear by the truth-tables of symbolic logic.

The name "hypothetical syllogism" is also given to two other types of argument whose major premiss is also a complex sentence:

> Queen Elizabeth II will go to Canada in 1985 or Prince Philip will go to Australia.
> But, Queen Elizabeth II will not go to Canada.
> So, Prince Philip will go to Australia.

Traditionally, this was known as a *modus tollendo ponens*. It has a negative minor premiss but an affirmative conclusion. Observe that nothing would follow if the second premiss had been the affirmation that Queen Elizabeth II will go to Canada, because the two alternatives are compatible. A "not both" connector requires an affirmative minor premiss to produce a negative conclusion.

> Ronald Reagan and Gus Hall will not both be elected President of the United States in 1984.
> But (let us say), Ronald Reagan will be elected.
> So, Gus Hall will not be elected President in 1984.

It would not prove anything to deny that Gus Hall will be elected as our minor premiss, because the negation of Hall's election is compatible with the negation of Reagan's election. The major premiss allows both parts to be false--as in a Mondale victory--but not both parts to be true. We now have four forms:

modus ponendo ponens	*modus tollendo tollens*
If p then q	If p then q
But p	But not q
Thus q	Thus not p

modus tollendo ponens	*modus ponendo tollens*
p or q	Not both p and q
But not p	But p
Thus q	Thus not q

Exercise: Please state whether the following are valid or invalid. If invalid, why; if valid, according to which rule?
Sample:
> If God exists, Cardinal Richelieu has been punished.
> God exists.
> Therefore, Cardinal Richelieu has been punished.
> Valid.
> *Modus ponendo ponens*.

1. You cannot serve both God and Mammon.
   John does not serve God.
   Therefore, John serves Mammon.
2. If this substance is lead, it will bend easily.
   It does not bend easily.
   So, it is lead.
3. You cannot have your cake and eat it.

You have eaten your cake.
So, you cannot have your cake.
4. Mary's mother or Mary's father was born in China.
Mary's mother was born in China.
So, Mary's father was not born in China.
5. If this is solid iron, it will sink in water.
It is solid iron.
So, it will sink in water.
6. Hegel or Kant taught at the University of Berlin.
Kant did not teach at Berlin.
So, Hegel did.
7. If this is dry pine wood, it will burn fast.
This does not burn fast.
So, it is not dry pine wood.
8. If this is copper, it will conduct electricity.
This is not copper.
So, it will not conduct electricity.

2.34   Truth-Tables

It will be helpful for us to review the conditional as presented in Section 1.25.

Consider the statement: "If this is Belgium, it must be Tuesday". It consists of two atomic propositions: "This is Belgium" and "Today is Tuesday", linked by the "if...then..." relation. Note that the molecular statement does not say that it is Tuesday if and only if it is Belgium, because Tuesday will keep recurring. There are four conceivable combinations of events, depending on whether it is Belgium or not, and whether it is Tuesday or not. Now, the molecular proposition--the conditional as a whole--is compatible with both of the atoms being false--it allows for it being Wednesday and Luxemburg, or with only the consequent true--it might be Rumania and Tuesday again--and, of course, for the possibility that it is indeed Tuesday in Belgium. What the conditional sentence excludes is arrival in Belgium on another day. We can write this as a table, where T means true and F false.

It is Belgium	It is Tuesday	If it is Belgium, it is Tuesday
T	T	T
T	F	F
F	T	T
F	F	T

The difference between this statement and the more drastic "It is Belgium, if and only if it is Tuesday" (which would condemn one to be in Belgium once a week) can be expressed in the line, where the first atomic sentence is false and the second true. The molecule is now false. Equivalence, as we noted in the section on immediate inference, means that the two propositions are true together or false together. So,

It is Belgium	It is Tuesday	It is Belgium, if and only if it is Tuesday
T	T	T
T	F	F
F	T	F
F	F	T

"It is Belgium and it is Tuesday" would be true if and only if both atoms are true. The table, therefore, is:

It is Belgium	It is Tuesday	It is Belgium and it is Tuesday
T	T	T
T	F	F
F	T	F
F	F	F

"Or" in English is equivocal; it has two meanings. It may mean a non-exclusive pair of alternatives (*vel* in Latin), as when one is asked whether one wants cream or sugar with coffee; "yes" means either or both. By contrast, a well-mannered person, asked whether he or she desires milk or lemon with tea or coffee, recognizes an exclusive "or" (*aut* in Latin), where "yes" means one or the other. Occasionally, one means one or the other or neither, as in "You will read this book or flunk the course". Considering the ambiguities of language, we must now distinguish between three different tables for "or" as opposed to the single table presented in Section 1.24. We thus have three tables (where Q and R stand for the atomic propositions:

Q	R	disjunction one or both (*vel*)	exclusive or, alternation either or (*aut*) at most one
T	T	T	F
T	F	T	T
F	T	T	T
F	F	F	F

Exercises: Write the truth-tables for:
1. If Blakeley gets a walk and Colbert hits a single, then Satty will hit a home run.
2. Satty will hit a home run or Colbert will not hit a single.
3. Blakeley gets a walk or he does not.
4. Colbert hits a single and fails to hit a single.
5. Satty will hit a home run if and only if Blakeley gets a walk or Colbert hits a single.

2.35   Formal Systems

Some logicians--most notably Whitehead and Russell in their *Principia Mathematica*--have organized their work into an axiomatic system. An axiomatic system has as its fundamental goal economy of thought. Everything depends on a minimum number (at least one each) of undefined notions and of primitive statements or axioms. There must also be two meta-rules which are not exactly in the system but rather about it: a rule for defining new terms from old one(s); and a rule for inferring theorems from axiom(s).

Much work has been done on trying to find ways to establish the consistency of logical systems as well as to find a mechanism for deciding the logical consistency or validity as the case may be, of all the propositions expressible in the language of a system. At a fairly elementary level, the truth-tables mentioned in the preceding section offer one such decision procedure.

The logic of terms presented above has been formalized with the aid of set theory and of the theory of functions. Key notions in the theory of sets were presented above (Section 1.1).

The formalisms of contemporary symbolic logic are extremely powerful tools. In a few cases, however, they are counter-intuitive. Thus, it is customary to translate "All A are B" as a hypothetical statement:

If x belongs to set A, then x belongs to set B.

This has an odd result in at least one foreseeable case. The proposition "All gobblins are purple" would be true on this analysis; for, since there are no gobblins, the case never occurs when the antecedent is true and the consequent is false. Such is the case with the "unicorn" in Section 1.3. But, it would also be true that "All gobblins are green", and so forth. This is not a philosophic problem as long as we understand that we have chosen for clarity a symbolism which does not correspond in a limit-situation to our common-sense intuition. That certainly does not prove that our intuition is wrong.

2.351 Tautologies: Truth-Table Verification

Logic has a rudimentary but vital syntax. We have already alluded to Russell's theory of types which requires that one carefully observe a hierarchy of elements, classes, and classes of classes, and that one only use a certain type of argument for a certain type of function.

For example, "x smokes a pipe" or "y dances the tango" are expressions that call for the name of an individual to be inserted in place of the variable. This name is called an "argument". Thus, we might substitute "Sherlock Holmes" for x and "Evita Peron" for y. Of course, syntax does not deal with making things true, but with making them meaningful, at least in so far as correctness contributes to meaningfulness. Therefore, we might substitute "Evita Peron" for x and "Sherlock Holmes" for y and get perfectly good, although doubtfully true propositions--"Evita Peron smokes a pipe" and "Sherlock Holmes dances the tango". What logical syntax will not allow, for example, is that we substitute a predicate function in a place where an individual argument is called for; we cannot say "Dances a tango smokes a pipe" for "dances a tango" is not the name of an individual.

The simplicity of propositional logic permits precision and

completeness in the exposition of the concept of "well formed formula".
Three rules define what is a proposition or well formed formula:

1. A proposition is a simple (or "atomic") statement.
2. A proposition is also the negation of such a statement.
3. A proposition is also the alternation, conjunction, implication nexus of two statements.

These rules apply to each other and to themselves. Thus, the disjunction of two negatives or the negation of a disjunction or a compound (molecular) hypothetical--in which the antecedent is a conjunction and the consequent an alternation--or, finally, the negation of a negation of a negation--all are examples of propositions as defined by these rules.

There are two kinds of functors whose arguments are propositions. They are monadic (having one argument) and dyadic (having two). It would not be impossible for there to be other levels of connectives, but it is neither necessary nor useful. Negation is the only monadic functor. Conjunction, alternation, implication, and so on, are all dyadic. In molecular propositions it is customary to use parentheses to make clear what is the argument of what. Thus, ~pVq is not the same as ~(pVq). In the first case the argument of ~ is p, whereas in the second it is (pVq). The first statement would read "Not-p or q", whereas the second would be "It is not true that p or q".

Again, the following are three different propositions: $p{\rightarrow}((p{\cdot}q)Vr)$ is not the same as $(p{\rightarrow}p){\cdot}(qVr)$ or $((p{\rightarrow}p){\cdot}q)Vr$.

A dyadic functor has exactly two arguments. "pV " and "pVqrs" are meaningless.

Two different systems have been worked out for the connectors of propositional logic. They are the Peano-Russell symbolism (which is identical in function to that of Hilbert, frequently used by mathematicians and basic to ours in this book) and the so-called Polish notation, invented by Jan Łukasiewicz.

In Peano-Russell notation, the negative functor precedes its argument, while the other functors occur between their arguments. This is a graphic way to portray their dyadic character. A slightly less graphic but in some respects simpler and more elegant procedure is that of Łukasiewicz, all of whose functors precede their arguments and no parentheses or points are needed. So,

Cpq	means "p implies q"	p→q
Kpq	means (the conjunction of) p and q	p•q
Apq	means (the alternation of) p or q	pVq
Np	means not-p, the negation of p	~p

Of course, one cannot say Cp or CPQRS in Łukasiewicz. Used correctly, however, each of his formulae means something quite distinct.

CpAqr	is "p implies the alternation 'q or r'"	p→(qVr)
CApqr	is "the alternation 'p or q' implies r"	(pVq)→r
ACpqr	is "the alternatives are that 'p implies q' or r"	(p→q)Vr

These brief remarks should reinforce the work of the text and explain some of the symbolism of the appendices (where we return to Polish notation). They also help to articulate some further notions. Once we have required that a proposition be a well formed formula, we can go on to say some other things about it:

A well formed formula can be consistent or not.

A well formed formula can be valid or not.

A consistent proposition is true at least once; it may be true more often. A valid proposition is always, i.e. necessarily, true on logical grounds.

An inconsistent proposition is never true; i.e. it is impossible on merely logical grounds. An invalid proposition is sometimes false. A valid proposition is also consistent. An inconsistent proposition is also invalid.

pV~p	is valid
p	is consistent (but invalid)
p•~p	is inconsistent

The method of truth-tables is a quick and certain way of determining consistency, validity, and their opposites.

The tautology is a type of proposition that is often met with in logic. A tautology is a proposition which is always true regardless of the truth values of its components. We can use truth-tables to establish the validity (truth in all instances) of a tautology.

A simple tautology is pV~p. This means that a proposition p is either true or it is false (a trivial notion, since this is the definition of a proposition). We can verify this tautology using the following truth-table.

Tautology 1: *Excluded Middle*   "A proposition must be either True or False. There is no alternative."

p	~p	pV~p
T	F	T
F	T	T

Column p contains all possible truth values for p. Column ~p contains the values of ~p according to the definition of negation with respect to p. Column pV~p contains the values of the disjunction according to the definition of disjunction. We notice that pV~p is true, independent of the truth value for p. Thus, we say that pV~p is a tautology.

Tautology 2: *Double Negation*

p	~p	~~p	~~p↔p
T	F	T	T
F	T	F	T

Another simple but important tautology is ~~p↔p. This says that the negation of the negation of a proposition is equivalent to (has the same truth value as) the original proposition. We understand the biconditional to be the "is" or "is equal to" in propositional logic. Thus, this tautology permits us to substitute a proposition for its double negation.

So, we see that p and ~~p are always either true at the same time or false at the same time; so, ~~p↔p is a tautology.

The next four tautologies are not as obvious as the above two, but are as important. By employing the next four tautologies we will discover that the symbols ~, • , V, → and ↔ are three more than we need. In other words, we will discover that all of the propositional calculus can be done using ~ and any one of the other four symbols.

Tautology 3: *DeMorgan's First Law* "The negation of an 'or' is the 'and' of the negations"

$$\sim(pVq) \leftrightarrow (\sim p\bullet\sim q)$$

p	q	pVq	~(pVq)	~p	~q	~p•~q	↔
T	T	T	F	F	F	F	T
T	F	T	F	F	T	F	T
F	T	T	F	T	F	F	T
F	F	F	T	T	T	T	T

Notice in this table that we need four rows instead of two because we must account for every possible combination of truth values of p and q. Once the values of p and q are established, the remainder of the truth-table is determined by the definitions of ~ , • , and V.

Tautology 4: *DeMorgan's Second Law* "The negation of an 'and' is the 'or' of the negations"

$$\sim(p\bullet q) \leftrightarrow (\sim pV\sim q)$$

p	q	p•q	~(p•q)	~p	~q	~pV~q	↔
T	T	T	F	F	F	F	T
T	F	F	T	F	T	T	T
F	T	F	T	T	F	T	T
F	F	F	T	T	T	T	T

Tautologies 3 and 4 are both Laws of DeMorgan, and we actually can prove 4 using 3 and double negation.

Derivation of DeMorgan 4 from DeMorgan 3 and Double Negation
We know that

(1) $\sim(pVq) \leftrightarrow (\sim p\bullet\sim q)$

by DeMorgan 3. Since p and q are any propositions whatsoever, we can let p=~r and q=~s without loss of generality. Substituting into (1), we get

(2) $\sim(\sim r \ V \sim s) \leftrightarrow (\sim\sim r\bullet\sim\sim s)$

From double negation,

(3) ~(~r V ~s) ↔ (r•s)

We can negate both sides, since when p↔q is true, so too is ~p↔~q, from the definition of the biconditional. So,

(4) ~~(~r V ~s) ↔ ~(r•s)   by double negation,
(5) (~r V ~s) ↔ ~(r•s)

which is equivalent to DeMorgan 4.

There are also other laws of DeMorgan, all of which are derivable from our theorems.

Tautology 5: *Material Implication*

$$(p{\rightarrow}q) \leftrightarrow (~pVq)$$

p	q	p→q	~p	~p Vq	(p→q) ↔ (pVq)
T	T	T	F	T	T
T	F	F	F	F	T
F	T	T	T	T	T
F	F	T	T	T	T

This tautology permits us to substitute ~pVq whenever we have p→q. (Note that this tautology verifies the computer circuit representation of implication in Section 1.4). Although this is convenient, it is a bit strange in more advanced logical work, for the implication has a meaning which differs from disjunction. This distinction is more apparent when we use the language of sets and understand that subset is defined classwise while union was defined elementwise (see 1.14 and 1.2). We will not be concerned here with this rather sophisticated problem.

Tautology 6: *The Biconditional* (and how it got its name)

$$(p \leftrightarrow q) \leftrightarrow [(p \rightarrow q) \cdot (q \rightarrow p)]$$

p	q	p↔q	p→q	q→p	•	↔
T	T	T	T	T	T	T
T	F	F	F	T	F	T
F	T	F	T	F	F	T
F	F	T	T	T	T	T

We can now see that the biconditional is just that: the conditional (implication) in both directions. Using tautologies 5 and 6, we have verified the computer circuit representation of the biconditional.

Tautology 7: *Substitution*

$$[(p \leftrightarrow q) \cdot (q \leftrightarrow r)] \rightarrow (p \leftrightarrow r)$$

p	q	r	p↔q	q↔r	•	p↔r	→
T	T	T	T	T	T	T	T
T	T	F	T	F	F	F	T
T	F	T	F	F	F	T	T
T	F	F	F	T	F	F	T
F	T	T	F	T	F	F	T
F	T	F	F	F	F	T	T
F	F	T	T	F	F	F	T
F	F	F	T	T	T	T	T

From the above tautologies we can demonstrate how all truth functions can be written using only ~ and V :

	FOR	USE
(i)	p→q	~pVq
(ii)	p•q	~(~pV~q)
(iii)	p↔q	~[~(~pVq)V~(~qVp)]

The first one (i), is simply tautology 5.

The second one (ii) is tautology 3 (DeMorgan's Law), along with tautology 2, double negation, and Tautology 7, substitution.

~(~pV~q)↔~~~p•~~~q	(Law of DeMorgan 3)
(~~p•~~~q)↔p•q	(Double Negation)
Thus, ~(~pV~q)↔p•q	(Substitution)

The third (iii) is the most complicated of them all and is a mixture of tautologies 2-7 as we may follow below:

~[~(~pVq) V ~(~qVp)]	
↔~[(~~p•~q)V(~~q•~p)]	(DeMorgan 3)
↔~[(p•~q)V(q•~p)]	(Double Negation 2)
↔[~(p•~q)•~(q•~p)]	(DeMorgan 3)
↔(~pVq)•(~qVp)	(DeMorgan 4 and Double Negation)
↔(p→q)•(q→p)	(Implication 5)
↔(p↔q)	(Biconditional 6)
QED!	

These two proofs are called "axiomatic deconstructions" and it should be noted that substitution is employed continuously.

Two further truth-table verifications are those of *modus ponendo ponens* and *modus tollendo tollens* as presented in Section 2.33. Symbolically, we have:

$$[(p→q)•p]→q \qquad \textit{modus ponendo ponens}$$

p	q	p→q	(p→q)•p	[(p→q)•p]→q
T	T	T	T	T
T	F	F	F	T
F	T	T	F	T
F	F	T	F	T

$$[(p\rightarrow q)\cdot\sim q]\rightarrow\ \sim p \qquad \textit{modus tollendo tollens}$$

p	q	p→q	~q	(p→q)•~q	~p	[(p→q)•~q]→ ~p
T	T	T	F	F	F	T
T	F	F	T	F	F	T
F	T	T	F	F	T	T
F	F	T	T	T	T	T

Note that *modus tollendo tollens* is a result of *modus ponendo ponens* on the law of contraposition:

$$(p\rightarrow q)\leftrightarrow(\sim q\rightarrow\sim p)$$

Exercises:
1) Define V, → and ↔, using ~ and • only.
2) Define • , V , ↔, using ~ and →.
3) Define • , V , → , using ~ and ↔ .
4) Prove your definitions correct, using an axiomatic deconstruction.
5) Prove the law of contraposition:
   $(p\rightarrow q)\leftrightarrow(\sim q\rightarrow\sim p)$, using truth-tables.

*   *   *

There are an indefinite number of theorems that can be derived from the basic axioms of an axiomatic system, some of which are listed in Appendix 2.

Moving from mainly implications to mainly alternations to mainly conjunctions could seem to be just interesting or "merely a game". On reflection, however, we can see that, although the ordinary language used by most members of a culture is relatively stable and homogeneous, the language of special groups--especially of various professions--is marked not only by greater frequency of certain words (and expressions) but also by a penchant for certain modes of argumentation and a predominance of certain logical forms. For instance, a lawyer will use the term "evidence" where a doctor uses the term "symptom", and "judge" rather than "decide"; but also the lawyer will tend to reason from "consequences" ("if the suspect had opportunity and motivation then..."), while the medical mind will work rather in terms of a concatenation or conjunction of distinct symptoms (such a temperature, with such a skin color, with such spots on the white of the eyes). The political speaker will use alternation because he is largely involved in the contentious presentation of alternative policies, interests, etc. We will come back to this point on a practical plane when we deal with poly-categoriality and semantic parsers.

Exercises: Do truth-tables for the following and say whether they are valid or not, and consistent or not.

1. ~(p•q)↔ (~pV~q)
2. ~(pVq)↔ (~p•~q)
3. (p→q)Vp
4. (p→q)→(p•q)
5. (pVq)•(~p•~q)
6. (p→q)↔(q→p)
7. [(pVp)→ p]•p

2.36   Computer Science: NAND and NOR Gates

In Section 1.4 we spoke of AND, OR, and NOT gates in computer circuitry. Now we can speak of two other gates which are used--NAND and NOR. These are the two most important gates, because any kind of gate can be formed by either gate. The truth-tables for these gates are:

p	q	p NAND q		p	q	p NOR q
T	T	F		T	T	F
T	F	T		T	F	F
F	T	T		F	T	F
F	F	T		F	F	T

Upon inspection, we may notice that p NAND q↔~(p•q) and p NOR q↔~(pVq). That is, NAND means not-and and NOR means not-or.

As circuits we have:

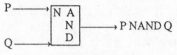

→ P NAND Q   where current flows out only
when no current is input at
at least one input

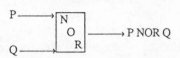

→ P NOR Q   where current flows out only
when no current is put in
from either input or both

Although it seems we should need two operations in order to get any other operation (as in Section 1.4), the fact that NOR (or NAND) is already a combination of two operations satisfies the requirements. Of course, we need to "trick" the circuit into acting as we wish. For example, we can make a NOR into a NOT as follows:

When *both* inputs are p, the result is necessarily ~p. Since both inputs are p, they both have the same value; so, when both inputs are true, the output is false (first line of the truth-table), and when both inputs p are false the output is true (last line of the truth-table). No other combination is possible. Thus, we have used NOR to emulate a NOT gate:

(1)   (p NOR p)↔~p

Using two NOR gates, we can emulate an OR gate as follows:

In other words,

(2) [(p NOR q) NOR (p NOR q)]↔ p∨q

We can use truth-tables to illustrate relations (1) and (2) above:

126

(1)

p	p NOR p	~p	↔
T	F	F	T
F	T	T	T

(2)

p	q	p NOR q	(p NOR q) NOR (p NOR q)	pVq	↔
T	T	F	T	T	T
T	F	F	T	T	T
F	T	F	T	T	T
F	F	T	F	F	T

The remaining relations of →, •, ↔ , NAND follow from the results of Section 2.35. For example, since

$$(p{\to}q){\leftrightarrow} \sim pVq$$

and

$$(\sim pVq){\leftrightarrow}\{[(p \text{ NOR } p)\text{NOR } q] \text{ NOR } [(p \text{ NOR } p) \text{ NOR } q]\}$$

then, p→q can be represented by:

$$[(p \text{ NOR } p) \text{ NOR } q] \text{ NOR } [(p \text{ NOR } p) \text{ NOR } q]$$

The other relations follow from Section 2.35 as:

(3) $p{\bullet}q{\leftrightarrow}\sim(\sim pV{\sim}q){\leftrightarrow}[(p \text{ NOR } p) \text{ NOR } (q \text{ NOR } q)]$

(4) $(p{\leftrightarrow}q){\leftrightarrow}\sim[\sim(\sim pVq)V{\sim}(\sim qVp)]{\leftrightarrow}[(p \text{ NOR } p) \text{ NOR } q]$
     $\text{NOR } [(q \text{ NOR } q) \text{ NOR } p])$

Exercises:
1. (a) Use truth-tables to show that (3) and (4) are
        tautologies.
    (b) Draw NOR gate circuits for p•q and p↔q.
2. (a) Find expressions equivalent to ~ , • , V , → and ↔ ,
        using NAND only.
    (b) Using tautologies, prove (a).
    (c) Draw NAND circuits for the relationships in (a).
3. Find NAND and NOR expressions for V (exclusive or) and

2.4   Induction

One of the most controversial topics in logic today is induction. Induction seems to be the opposite of the kind of formal or deductive reasoning we have been studying, going from the particular to the general, instead of from the general to the particular. That in itself might make us wary! Induction involves generalization. When we know enough members of a type, we affirm that all the members of that type have the properties we have found. For example, having discovered that all the dogs we know wag their tails when they are happy, we conclude: "All dogs whatsoever wag their tails when they are happy". This is known as induction by simple enumeration. The enumeration is incomplete because we do not examine all the dogs in the world. As most people realize, it is more useful to examine different breeds of dog in different situations than an equal or even much larger number of dogs of the same breed in identical situations. This type of induction can be done on a common-sense level spontaneously, although it has some applications also in the careful classifications of the biological sciences.

There is an important difference between the following apparently isomorphic inductions. *First*, "Snoopy barks. Lassie barks. Rin Tin Tin barks. Benji barks. ... Therefore, all dogs bark." This is an inductive leap from some (a sufficient but incomplete number) to a universal statement describing a state of affairs. *Second*: "Pigs have lungs. Dolphins have lungs. Bats have lungs. Cats have lungs. ... So, all mammals have lungs." In the second case, we are enumerating species of the genus (in logical if not biological terminology) "mammal". It is possible to do a complete enumeration of those species, since we can inspect each kind of mammal, which is not the case for individual dogs, since we cannot inspect every dog.

More problematic are the inductions that lead to statements of causal laws. The classical description of this type of induction was given by John Stuart Mill. By a causal law, Mill understood a regular temporal sequence of events. If B regularly follows A, then A is said to cause B. Mill envisaged sifting out data according to various patterns or "canons". For example, we can study a number of cases of some phenomenon we need to explain--let us say, juvenile delinquency--and, if we are lucky, we discover something common in the antecedents of each case--let us say a broken home. Notice that even this oversimplified example does not suggest that all broken homes produce juvenile delinquents. This method of presences or similarities gives us a "necessary" cause, i.e., a broken home is necessary for a juvenile delinquent to come to be.

The second approach within this method is to study cases, where what we want to explain is not present, although it might be. Let us say, for the sake of example, that we now study teenagers who are not juvenile delinquents and discover that none of them uses drugs. There is something absent in all the cases where the effect we wish to explain is absent. We now conclude that drug use is a "sufficient" cause of juvenile delinquency. It may not be the only cause; it may not be present in all cases, even in combination with something else (i.e., be a necessary cause). That is, drug use is not a necessary cause for juvenile delinquency, as a broken home is not sufficient.

Observe that this is a simple method for excising coincidences. Let us suppose that the first tropical storm of the season--call it "Alex"--was preceded by one of Fidel Castro's marathon speeches. If we use the method of presences we will examine the antecedents of hurricanes Bertram, Carl, Dennis, Edgar, etc., and doubtless find that others were not preceded by oratory. If we use the method of absences or inverse method of presences (here, good weather), we will doubtless find some nice days preceded by windy discourse.

Necessary and sufficient causes sound metaphysical. In the last analysis, of course, if we did not believe that the world, at least in its material aspects, is really governed by determinate laws, looking for them makes little sense. John Stuart Mill's intention, however, was to avoid metaphysical entanglements altogether and to offer a methodologically useful distinction. Indeed, the notions of necessary and sufficient have a merely logical use, far from the world of real causality, in mathematics. It is necessary that a number be an integer for it to be even. It is sufficient that it be divisible by four or be an integer greater than five but less than seven for it to be even. It is necessary and sufficient that it be divisible by two. By putting our solution in hypothetical form, i.e., "If...then...", we can refer to Section 1.25 for the interpretation of necessary and sufficient:

1. If a teen uses drugs then he will be a juvenile delinquent (Drugs are sufficient to generate juvenile delinquency).
2. If a teen is a juvenile delinquent then he came from a broken home (a broken home is necessary for juvenile delinquency).

Note that in 1. we are not saying that all juvenile delinquents use drugs but instead that all teenage drug users are juvenile delinquents. One need not use drugs to be a juvenile delinquent--but if one did, one would be.

In 2. we are not saying that all broken homes cause juvenile delinquency, but instead that all juvenile delinquents must have come from broken homes. That is, broken homes may cause straight-arrows as well as juvenile delinquents, but a juvenile delinquent could not have come from a non-broken home.

Mill's solution was to define these terms temporally: hence, that which always precedes a given effect is a necessary cause, whereas that which is always followed by a given effect is a sufficient cause.

If it turns out that if we find something present in the antecedents when the problem we study occurs, and absent when the problem does not occur, then we have identified a "necessary and sufficient" cause. Were it true that every teen that used drugs was a juvenile delinquent and that every juvenile deliquent used drugs, then drug use would be a necessary and sufficient cause for juvenile delinquency--and juvenile delinquency would be necessary and sufficient for teen drug use. Both teen drug use and juvenile delinquency would have to occur if either were to occur.

For Mill the method of presences and the method of absences were two versions of the same format, the canon of agreement in which one sought amid otherwise very varied circumstances some common factor (positive or negative). Mill's canon of difference must be carried out under almost laboratory conditions, because ideally we seek to draw a connection between a single factor which is present when the phenomenon to between a single factor which is present when the phenomenon to be explained occurs and absent when the phenomenon to be explained fails to occur. All other circumstances should be the same.

Mill mentions other methods or canons of induction. One of them, that of concomitant variation, involves two sets of factors, that can be present in varying degrees and that increase or decrease together. This type of induction led the Surgeon General of the United States to conclude that smoking causes lung disease. We must have two factors that do not simply occur or not occur but happen to a greater or lesser extent. Thus, a person can smoke more or less and within a population at a given level of tobacco consumption, there can be greater or lesser incidence of lung trouble. This type of relation may also work negatively. "Driver education prevents accidents" would depend on the same type of correlation between amounts of instruction and population. Evidently, in an individual case, there are perfect drivers who need no formal instruction and thoroughly indoctrinated youths who are accident-prone. Also, there may be limits to the concomitant variation, beyond which a kind of diminishing return takes over. Proof that someone had completed ten semesters of driver education would probably not prove superb ability but hopeless ineptitude. Finally, Mill's method of residues uses a kind of process of elimination, whereby once we show a complex cause of a complex event, we then identify specific antecedent factors as leading to specific factors in the consequent.

The outer planets of our solar system were discovered by something like this method. The behavior of Uranus (discovered in 1781) was not sufficiently explained by known masses and the presence of undiscovered planets was predicted. This led first to the discovery of Neptune (1846) and then to the discovery of Pluto (1930). These canons have a limited use as rules of thumb. They contain all sorts of theoretically questionable assumptions. The very notion of causality as a sequence--one thing after another--is flawed. Obviously, when I say my writing causes words to appear on the paper, I do not mean that I write now and see words appear later! The temporal analysis fails to distinguish between condition and cause; in fact, it more clearly applies to condition than to cause. The first three canons assume that we know what the relevant antecedent factors are. They might be workable in a familiar situation; e.g. used by the school physician in the face of an epidemic of stomach-aches to find out what was wrong with the cafeteria food; but would be less useful in genuine discovery where the problem is to see for the first time that something is relevant, i.e. that mosquitoes have something to do with malaria, as opposed to the unhealthy night air, already blamed by the Romans. The method of concomitant variation betrays the fact that we do not fully understand the causal mechanisms involved, as the smoking example shows; and the method of residues assumes that individual antecedents can be aligned with individual factors in the consequent which, e.g., in chemical reactions, is certainly not true. Furthermore, the reasoning in the method of residues is more deductive than inductive; we are guided by an over-arching law (gravity, in our example), proven or at least assumed as a hypothesis. Finally, as our simplistic examples suggest, reality may resist neat pigeon-holing.

A difficulty in understanding the nature of induction is often due to an incomplete differentiation between deduction and induction. It is frequently said that deduction moves from the general (or universal) to the particular (or individual or singular), while induction rises from particular to general.

The formal-structural difference between deduction and induction can be represented as follows:

Deduction:

> if, if p then q,
> <u>and p</u>
> then q

Affirming the antecedent grounds the affirmation of the consequent.

Induction:

> if, if p then q,
>> and q
>> then p

Assertion of the consequent leads to assertion of the antecedent.

Clearly, deduction's validity is due to the fact that the conclusion (consequent) is contained in its entirety in the antecedent. Induction, on the other hand, needs a consequent grounded by something other than the antecedent, in order to be able itself to ground the antecedent. We may gain further insight by viewing the truth-table of implication with respect to deduction and induction. In each table, p→q as true is encircled, while p is false is starred in deduction (since p is true) and q is false is starred in induction (since q is assumed true):

DEDUCTION				INDUCTION		
p	q	p→q		p	q	p→q
(T	T	T)		(T	T	T)
T	F	F		*T	*F*	F
*(F	T	T)		(F	T	T)
*(F	F	T)		*(F	*F*	T)
(only one possibility)				(two possibilities)		

Notice that in the deduction table the only possibility of p→q and p as both true is the top line, where q is true. But, in the induction table, both the first and third lines are possible interpretations for p→q and q true. So p can be true or false; which means there is no necessity for p being true in induction. In effect, the consequent of an induction needs a deductive grounding before it can, in turn, ground the antecedent.

The matter is even more complex when we ask what sort of proposition is represented by p and q.

If q represents the so-called "protocol sentences" (p-sentences) which express a discrete event taking place at a given place and time under specific conditions, then we have the classic case of "verification". Each p-sentence contributes to the probability of the antecedent.

If q represents a statement which itself results from an induction, then we have "explanation"; and, since the first-level induction was only probable, the second-level is doubly probable; therefore, it needs considerably more evidence.

Our notion of the tree can come to our assistance here, to illustrate the differences among the forms of deduction and induction. Here is the hypothetico-deductive tree in its simplest form:

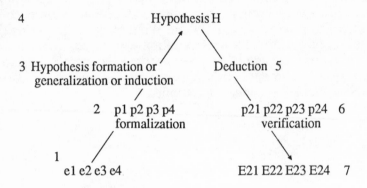

Comments:
  (1) We begin with events that result from observation and experiment, where we are able to specify time, place, state, etc., in
  (2) the so-called protocol sentences, each of which describes one and only one event;
  (3) generalization moves from several of these p-sentences to one H-sentence,
  (4) that restates them in synthetic, generalized form, so that

if p(1) and p(2) and p(3) and p(4), then H

  (5) the deductive phase has the form

if H, then p(23), p(24), p(25), ...

(6) Verification consists in confirming the p-sentences inferred from H by establishing (through observation and
experiment) that each p does describe an E.

Example:

E(1): a piece of phosphorus ignites at 60 C°
E(2): a piece of phosphorus ignites at 60 C°
E(3): a piece of phosphorus ignites at 60 C°
E(4): a piece of phosphorus ignites at 60 C°

and so on.

p(1) = "piece 1 of phosphorus ignites at 60 C°"
p(2) = "piece 2 of phosphorus ignites at 60 C°"
p(3) = "piece 3 of phosphorus ignites at 60 C°"
p(4) = "piece 4 of phosphorus ignites at 60 C°"

and so on.

Generalizing:

if p(1), p(2), p(3), p(4) ... then H = "all pieces of
phosphorus ignite at 60 C°"

Deducing:

if H, then p(23) "piece 23 of phosphorus ignites at 60
C°"

Verifying:

if E(23) is the case, p(23) is true and H is more
probable.

Deducing:

if H, then p(24) "piece 24 of phosphorus ignites at 60
C°"

Verifying:

if E(24) is the case, p(24) is true, and H is more
probable, ..., and so on.

Question: Does H ever become "true" rather than "more probable"?

Answer: Yes and No.

If we have a p for every E, then the H becomes true. This is
a "closed induction". It is uninteresting because it is in effect
a form of deduction; for, all p's from the very first to the

very last can be inferred from it. Such is the case with an induction, leading to the fact that "all mammals have lungs" by enumeration of each type of mammal (pigs, dolphins, bats, etc.).

But, H can become only more probable if it is to remain a "fruitful hypothesis", i.e., one from which we can infer new knowledge, which is the induction performed to determine that all dogs wag their tails when they are happy.

* * *

Real cases are usually much more complicated. Let us take the hypothetico-deductive tree for the taking of the Bastille--a key event in the French Revolution.

"The Bastille was taken by a mob when the soldiers guarding it became ineffective because enough had been shot to reduce the fighting potential of the guard units" is a more specific and far more useful version of "The Bastille fell when the soldiers could no longer effectively defend it". This latter statement is almost a truism covering the case of the loss of men, but also shortage of powder, demoralization, or any militarily significant difficulty, including the inability of German mercenaries to understand French commands (as happened at a critical moment in Louis XVI's attempted flight). The expanded version has the following structure with four components:

q   "shot soldiers are ineffective"
p   "a certain number of soldiers are shot"
r   "guard units with shot soldiers lose their fighting potential"
s   "a Bastille guarded by units with diminished fighting potential is taken"

or

$$(p \cdot q \to r) \to s$$

Comments:
   (1) p here stands for "soldier 1 is shot", "soldier 2 is shot", etc.
   (2) q represents an H that results from a previous process of generalization from p-sentences of the

form "shot soldier 1 is ineffective", "shot soldier 2 is ineffective", etc.

(3) r corresponds to using q in a higher-level generalization, where "shot soldiers" appears as sufficient but not necessary condition of weakened fighting potential.

(4) s is an H of even higher generality.

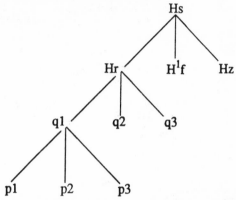

Exercise: Do something similar for "For want of a nail the shoe was lost; for want of the shoe the soldier was lost; for want of the soldier the squad was lost; for want of the squad the regiment was lost; for want of the regiment the army was lost".

* * *

Clearly, then, "deduction", "induction", "verification", "formalization", and other such terms take on various meanings depending on the level of generality. One can use the *Tree of Porphyry* for reference:

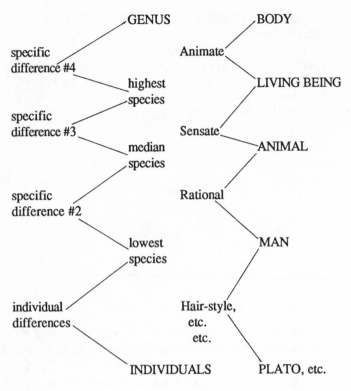

A deduction will be the more fruitful, the more subsidiary hypotheses (species) it illuminates. An induction is stronger, the more individuals it can include (verification) and/or the more generalizations (species) it can ground (explanation).

Since induction is involved in the process of establishing the premisses of an argument, it is clearly worth asking what sort of induction was involved in formulating a given premiss. A premiss that results from the more genuine form of induction ((1) above) has a probability that depends on the number of cases tested (which varies according to the type of experiment involved)--it is increased only theoretically, not empirically. The other forms of induction ((2) or (3) above)--from species to genus or from species to higher species--establish premisses, the probability of which is a complex function of both the theoretical links between species and species, and the empirical links among the cases that ground the most fundamental inductions involved in the process. This difference will affect how we deal with various language structures in Chapter III.

Exercises:
1. Describe briefly how one might organize an inductive study of the effect of logic courses on LSAT, GRE, or GMAT scores.
2. Can one do an inductive study to determine the cause of a historical event like the American Civil War? Why or why not?
3. Describe briefly how one might inductively study the relation of salt to heart trouble.
4. Describe briefly how you might try to discover the cause of stomach ulcers.

III
Language Analyzers

The art of manipulating valid forms is applied today in the construction of automatic language analyzers. Now that we have the elements and principles of logic in hand, let us look at the main types of these automatic language analyzers, and use our knowledge of set theory and propositional calculus to take steps toward constructing a working model.

Such analyzers can be syntactic, quasi-semantic, fully semantic; and, it turns out that all of them are needed for achievement of the final goal--conversational interaction of human with machine (which, incidentally, is also "human" but only indirectly so!)

3.1   The Syntactic Check

A syntactic checker pays attention to how the words in a sentence correctly relate to each other and, therefore, it makes extensive use of grammatical rules and categories.

As we saw at the end of Chapter I, for the computer a word is a "string" or ordered set of characters. In fact, for syntactic purposes it is useful to look at a sentence as a set, whereupon its words become its proper subsets, and the letters are the elements.

The earlier chapters have typically dealt with sentences in the standard form of Subject + Copula + Predicate. A true syntax checker should be able to take copulae as they are used by real speakers in actual speech situations. This is a prerequisite for solving the problems of artificial intelligence.

3.11   Sketch of a Syntax Checker

We begin with an input which is a complete simple sentence, made up of subject, copula, and predicate, and which ends with a period.

Since a compound sentence (two or more independent clauses linked by co-ordinating conjunctions) or complex sentence (one or more independent clauses linked by subordinating conjunctions with one or more dependent clauses) can be treated as "independent clause (complete sentence) plus X", we can take the complete simple sentence (independent clause) as basic input.

Our syntax checker must go through the following eight steps. In the remainder of this section, we will A) further explain these eight steps; B) demonstrate what is done in these steps with a concrete example, and C) design a computer program to accomplish the task of a syntax checker.

Step one: Enter a sentence. If nothing is entered, then prompt for input. If there is no terminal period or no initial capital letter, then prompt for correct input. Handle the terminal punctuation. (. , ! , ?)

Step two: Isolate the words. If there is an article (the, an, a), then ignore it, or preattach it (*word1* + *word2* = TERM), or flag it (because it is not needed for the syntactic phase). Alphabetize the words to speed up the search.

Step three: Locate the copula, using a "table look-up" (search a list of available copulae or verbs).

Step four: Define the subject as what occurs from the beginning of the sentence to the beginning of the copula; i.e.

SUBJECT = SENTENCE - (COPULA + COMPLEMENT).

Then, define the predicate as what remains after the bracketing of the subject and copula.

PREDICATE = SENTENCE - (SUBJECT + COPULA).

Step five: Verify that we have a valid subject of the form

(article) + (modifiers) + SUBSTANTIVE + (phrases)

In other words, there must be a noun (substantive) or something taken as a noun. There can but need not be an

141

article, modifier (adjective) and phrases (adjectival or adverbial).

Step six: Determine the type of copula [transitive or intransitive; passive or active; etc.].

Of course, the distinctions between transitive and intransitive, active and passive, and so on, are insufficient from a logical viewpoint, as we see from the basic structures of the following sentences:

subject + copula
   Johnny sleeps.
subject + copula + direct object
   Pam kisses Mike.
subject + copula + indirect object + direct object
   He gave them the money.
subject + copula + direct object + object-complement (adjective)
   The games made the children happy.
subject + copula + direct object + object-complement (noun)
   The vote made him president.
subject + copula + predicate-noun
   Emily was the pianist.
subject + copula + predicate-adjective
   Helga was very prompt.

Notice that the first of these and the last two are intransitive, while the rest are transitive.

A more interesting classification from a logical perspective is to point out that the first is monadic, the second and last two are dyadic, while the other three are triadic--where "monadic" means "having one argument", "dyadic" is "having two arguments", and triadic is "having three arguments". In "Johnny sleeps", "sleeps" is a monadic functor with one argument, "Johnny". In "Pam kisses Mike", "kisses" is a dyadic functor with two arguments, "Pam" and "Mike". When we say "He gave them the money", "gave" is a triadic functor, the arguments of which are "He", "them", and "the money". In the earlier example (see Section 2.1) "eggs" is monadic while "cooks" is dyadic.

Step seven: Certify that the predicate meets the exigencies of the copula.

For example, the transitive copula takes a direct object (noun) and possibly an indirect object (with or without preposition), while the intransitive copula will have after it something of the same status as the subject.

Step eight: Proceed to break down and verify the components of the subject and the components of the predicate--which involves detecting phrases, modifiers and their parts, and continues until all the words have been "located" relative to one another (again using table look-up).

```
10       ! THIS PROGRAM BREAKS SENTENCES UP INTO WORDS
20       ! S$(J) IS THE Jth SENTENCE
30       ! W$(K,J) IS THE Kth WORD IN THE Jth SENTENCE
40       ! TRIVIAL CHANGES WILL PERMIT INPUT FROM AND
45       ! OUTPUT INTO FILES.
50       ! LIMITS FOR NOW ARE 100 SENTENCES OF 10 WORDS.
60       ! LAST SENTENCE TO ENTER IS 'STOP' WITHOUT A
             PERIOD.
100      DIM S$(100),W$(10,100)
110      J=1
120      INPUT S$(J)
130      UNTIL S$(J)="STOP" !STOP IS FLAG TO INDICATE LAST
             SENTENCE
140      G$=CVT$(S$(J),16%+32%)
150      L = LEN(G$)
160      IF RIGHT$(G$,L)<>"." THEN PRINT "ENTER SENTENCE
             WITH PERIOD AT END"\GOTO 240      !CHECKS FOR
                                                 PERIOD
180      K=1
190          FOR I = 1 TO L
200              IF MID$(G$,I,1)='.'THEN W$(K,J)=LEFT$(G$,I-1)
                 \G$ =RIGHT$(G$,I+1)\K=K+1\I=1
202              IF MID$(G$,I,1)='.' THEN I=L
210          NEXT I
220      W$(K,J)=LEFT$(G$,LEN(G$)-1)  ! LAST WORD WITHOUT
                                         PERIOD
230      J=J+1         !PREPARE FOR NEW SENTENCE
240      INPUT S$(J)
600      LET P$=','
650          FOR H=1 TO J-1
```

```
700          PRINT W$(K,H);P$; FOR K=1 TO 10
730             PRINT
750          NEXT H
999     END
```

The work of the *automatic syntax checker* is done when each word can be immediately linked correctly to some other word and, through it, to all the other words in the sentence.

Example: "The animals fled the approaching fire."
Step one: Read the last character of the input.
   Is it a period? (cf. line 160 above).
   If not, ask for a new input.
   Does the first word begin with a capital letter?
   If not, ask for a new input.
Step two: Read the first word.
   Is "The" in the copula-table? No.
   Is "animals" in the copula-table? No.
   Is "fled" in the copula-table? Yes.
Step three: Define: subject = "The animals"
   (i.e. whatever precedes the copula)
   copula = "fled" (better: "were fleeing" or "= fleeing")
   predicate = "the approaching fire" (i.e. whatever follows the
       copula)
Step four: (verify the subject): Read the first word of the subject.
   Is "The" a noun? No.
   Is "animals" in the noun(substantive)-table?
   If not, ask for informational input.
   If yes, then we have a valid subject.
Step five: Parse the copula.
   Is "fled" transitive?
   If no, treat the predicate as complement (of time, place, etc.).
   If yes, use Step four to find the noun.
Step six: Certify that copula and predicate correlate one with
   the other.
   Is "the" the direct object? No.
   Is "approaching" the direct object? No.
   Is "fire" the direct object? Yes.
Step seven: Deal with the remaining terms.
   For example, "approaching" fits a simple rule that "what

144

precedes a noun and ends in "-ing" modifies that noun".

COPULA-TABLE			NOUN-TABLE	
EAT	ATE	EATEN (trans)	ANIMAL	ANIMALS
SEE	SAW	SEEN (trans)	DOG	DOGS
HAVE	HAD	HAD (trans)	CAT	CATS
GO	WENT	GONE	FIRE	FIRES
SEEM	SEEMED	SEEMED	BOOK	BOOKS
FLEE	FLED	FLED (trans)	ROAD	ROADS
SEIZE	SEIZED	SEIZED (trans)	FACT	FACTS
ACT	ACTED	ACTED	KNIFE	KNIVES
FEED	FED	FED (trans)		
etc.				

Clearly there will be a large number of tables. However, the speed of the computer is such that it can do several such look-ups per second. See Section 3.4 for the structure of the prompting and the organization of the files needed for both syntax and semantics.

Even at this elementary level, one can design a simple, if crude, conversational capability into the machine. One could, for example, set up a table of terms (dog, bark, canine, guard, tail-wagging, hunting, Doberman, etc., etc.) and instruct the machine to take the input word (e.g. "bark") and select one of its correlatives (e.g. "hunting") for use in an answer. Of course, this will be quite hit-and-miss, unless one is constantly refining the table in accordance with any nonsense produced.

For example, if we do an

```
500    OPEN "MARXTALK.DAT" FOR OUTPUT AS FILE #1%,
```

assuming that the basic terms of Marxian discourse include (but are not limited to) those across the top of the following table, we can initiate a simple give-and-take with the machine:

	need	production	value	accumulation	alienation	class confl.	revolution	dict.of prol.	socialism	Communism
substance	A$(1) lack	A$(2) worker	A$(3) labor	A$(4) wealth	A$(5) theft	A$(6) inequality	A$(7) justice	A$(8) united front	A$(9) nationalisation	A$(10) humanity
quality	A$(11) vital	A$(12) prod.for use	A$(13) use-value	A$(14) tools	A$(15) exploitation	A$(16) classes	A$(17) strikes	A$(18) expropriation	A$(19) construction	A$(20) virtue
quantity	A$(21) essentials	A$(22) prod.for exch.	A$(23) exch-value	A$(24) capital	A$(25) oppression	A$(26) masses	A$(27) people	A$(28) majority	A$(29) commodious	A$(30) prosperity
relation	A$(31) imbalance	A$(32) transformation	A$(33) equivalence	A$(34) exploitation	A$(35) self-negation	A$(36) conflict	A$(37) barricades	A$(38) purge	A$(39) emulation	A$(40) communality
where	A$(41) biological	A$(42) labor space	A$(43) commodity	A$(44) goods	A$(45) class-conflict	A$(46) society	A$(47) streets	A$(48) class society	A$(49) new world	A$(50) universe
when	A$(51) ever	A$(52) work-time	A$(53) pay-day	A$(54) balance-sheet	A$(55) devaluation	A$(56) tension	A$(57) crisis	A$(58) socialism	A$(59) post-capit.	A$(60) eternity
action	A$(61) pursue	A$(62) modification	A$(63) profit	A$(64) hoard	A$(65) reject	A$(66) conflict	A$(67) revolt	A$(68) oppress	A$(69) liberate	A$(70) symbiosis
passion	A$(71) inclination	A$(72) satisfaction	A$(73) benefit	A$(74) enrichment	A$(75) sadness	A$(76) refusal	A$(77) displacement	A$(78) equalization	A$(79) homogenizaim	A$(80) happiness
disposition	A$(81) lack	A$(82) betooledness	A$(83) profitable	A$(84) acquisition	A$(85) isolation	A$(86) suspicion	A$(87) contention	A$(88) egalitarianism	A$(89) fraternity	A$(90) co-ordination
habitus	A$(91) desire	A$(92) want	A$(93) like	A$(94) have	A$(95) lust	A$(96) repulsion	A$(97) aggression	A$(98) purification	A$(99) community	A$(100) love

To speed up the process, both lists (the words in the input sentence, and the 100 words in the table) can be alphabetized.

Adopting a question-and-answer format, we could have a rule such that "if the input sentence begins with an interrogative term and ends with a question mark, then the output sentence reads the line in the table corresponding to the interrogative term to find the subject of the response". Thus, "where" points the machine to line 5 in the table, "when" to line 6, "who" to line 1, and so on. (A correlative rule could formulate a question out of a declarative sentence that is input).

Our basic Answer-the-question-on-Marx algorithm could be:

For any input sentence containing A$(X), ending with "?" and where WORD1 points to line Y, formulate a sentence ending with a period and containing A$(X+1) or A$(X-1) or A$(X-9) or A$(X-10) or A$(X-11) or A$(X+9) or A$(X+10) or A$(X+11). Schematically:

A$(X-11)	A$(X-10)	A$(X-9)
A$(X-1)	A$(X)	A$(X+1)
A$(X+9)	A$(X+10)	A$(X+11)

We can distribute this in ring-form as follows:

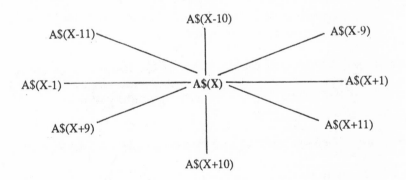

In other words, the terms available for formulating an answer stand immediately before, after, above and below the input term, as well as to its upper left and right, and lower left and right. We will refer to the elements which surround the given element as the "semantic ring" of relevance for element A$(X,Y). Note that the semantic ring is determined by adding or subtracting a "1" or 10 from either X or Y or both. Note also that the above table "wraps" around itself, so that the far right elements connect with those on the far left and the last with the first.

This "semantic ring of relevance" will be, of course, as valid and accurate as its formulator's grasp of the subject matter. What is more, unless the interlocutor is "on the same wave-length" as the formulator of the table, there is great risk of producing nonsense.

Using this table, a dialogue might run like this (between Alphonse, the computer, and Gaston, the human, with programming hints in parentheses):

Alphonse: "There are files ready for Aristotle, Plato, Alexander the Great, Napoleon, Priestley, Marx, and Superman. Which is to be discussed?"
Gaston: "Let us talk about logic."

(the program would search Gaston's answer to see if one of the appropriate names appeared. None did; therefore...)

Alphonse: "You must mention one of the following to initiate dialogue: Aristotle, Plato, Alexander the Great, Napoleon, Priestley, Marx, Superman. Which would you like?"
Gaston: "O.K. What about Marx?"

(500   OPEN "MARXTALK.DAT" FOR OUTPUT AS FILE #1%)

(note that the mere presence of "Marx" is enough to get things going. Even if Gaston had said "Dump Marx", the file would be opened)

Alphonse: "Marx' central notions include: need, production, value, accumulation, alienation, class-conflict, revolution, dictatorship-of-the-proletariat, socialism, Communism

(reading across the top of the table). Which interests
you?"
Gaston: "How many forms of value does Marx mention?"

"How many" + "?" points to line 3, where the algorithm
above locates "exchange-value" as A$(X), leading to:

A$(X-11)	A$(X-10)	A$(X-9)
production for use	use-value	tools
A$(12)	A$(13)	A$(14)
A$(X-1)	A$(X)	A$(X+1)
prod. for exch.	exch.-value	capital
A$(22)	A$(23)	A$(24)
A$(X+9)	A$(X+10)	A$(X+11)
transformation	equivalence	exploitation
A$(32)	A$(33)	A$(34)

Here is the ring:

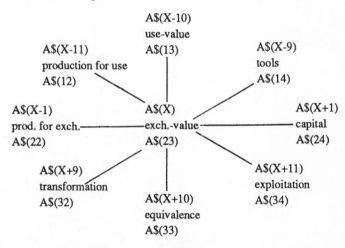

Alphonse: "Marx mentions 'exchange-value' and 'use-value',
although under 'value' he also mentions 'equivalence' and
the related notions of 'products-for-exchange', 'capital',
'tools', 'products-for-use', 'transformation', 'exploitation'."

(Note that our formulator forgot "absolute value", very key to Marxism!)

Gaston: "How does Marx explain exploitation?" ("How" points to line 4, where "exploitation" occurs)

(At this point "exploitation" becomes A$(X) and a new "semantic ring" grounds the question-and-answer process)

A$(X-11)	A$(X-10)	A$(X-9)
exch.-value	capital	oppression
A$(23)	A$(24)	A$(25)

A$(X-1)	A$(X)	A$(X+1)
equivalence	exploitation	self-negation
A$(33)	A$(34)	AS(35)

A$(X+9)	A$(X+10)	A$(X+11)
commodity	goods	class conflict
A$(43)	A$(44)	A$(45)

And, the ring is:

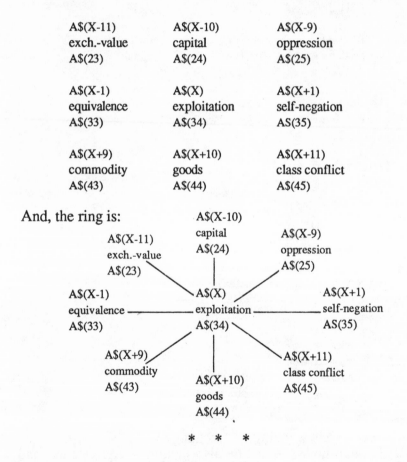

*   *   *

Clearly, such a give-and-take can go on indefinitely.

However, not only does it suffer the limitations we have already indicated, but further, if Gaston does not limit himself to simple functors like "mention", "discuss", etc, the whole exchange can quickly degenerate into "pidgin-Marx". Thus,

Gaston: "How does Marx see a majority being formed?"
Alphonse: "Marx sees a majority formed by people with commodiousness, expropriated in a purge."

Of course, these limitations can be reduced by:

1. expanding the semantic ring;
2. using interlocking semantic rings;
3. replacing the formulator (or "tutor");

but, it will be more expedient to equip Alphonse to be able to do a complete and adequate syntactic check and then to do a complex semantic categorization.

3.2 The Semi-Semantic Checker

The syntax checker assumes, as we saw, that we already have a grammar.

The semi-semantic checker takes up isolated ordering principles without attending to how they relate to each other. The semi-semantic checker is absorbing much of the attention of artificial intelligence research today. The main themes have to do with techniques for searching and sorting and with the construction of trees.

In elementary terms, a "search" is a process which finds a particular entry within a given set of possibilities. For example, in our previous conversation between Gaston and Alphonse, Gaston had to "search" through the set of Aristotle, Plato, Alexander the Great, Napoleon, Priestley, Marx and Superman for the entry "Marx" made by Alphonse. Actually, Gaston had to make two searches--one through Alphonse's sentence "O.K., what about Marx?", and the other through his set of legal possibilities--and find matching entries: "Marx" in both places.

Searches can be quite complicated unless the data set and the entered set are very small or conveniently arranged. Imagine, for example, trying to find the telephone number of Philip Smith. Unfortunately, he lives in Manhattan and the telephone books have a list of names and numbers in order of the date of telephone installation! (That is, the first person to get a telephone installed is the first name on the first page, and the last installation is the last name on the last page). Of course, if Philip Smith lived in Obediah, North Dakota--which has 25 telephones--you could find the number easily, independent of any ordering of numbers. Without any knowledge of when P. Smith had his telephone installed, you would have to search through every name until you found his name and corresponding number. It would make no difference where you started; all non-repetitive search procedures would depend on "luck" for a quick find. ("Non-repetitive" means you never read the same name twice).

Fortunately, telephone books are ordered alphabetically which makes the search much easier. Now, how do you search? Open the book about 2/3 of the way; find you are at "Richards"; turn about 75 pages forward; find you are at "Smyth"; turn back 15 pages; find you are at "Smith, J.". Turn forward 2 pages; find yourself between "Smith, Pe" and "Smith, Pi"; find "Smith, Philip"; find the proper address; find the telephone number; call; get busy signal; quit in disgust.

The way in which a set of data is ordered is called a *sort*. We are happy that telephone numbers are sorted alphabetically. Of course, the phone company needs other kinds of sorts for other purposes; for example, the installation sort. The way in which a set of data is sorted always affects the search. The easiest sort--random or unsorted data--produces the most time-consuming search. A more sophisticated sort--e.g. alphabetical order--produces a rather efficient binary search, as demonstrated in the previous paragraph. Obviously, we could always use a random or linear search but this would be rather foolish and unnecessarily time-consuming if we knew that our data were organized alphabetically.

For our purposes, we will be most interested in three different kinds of sort: categorial, frequency, and alphabetical. First, our files will be distinguished categorially--e.g. by parts of speech. We must separate nouns, verbs, adjectives, etc., and place them in separate files. Some of these files will then be separated into two distinct files, such as CACHENOUN and NOUN, where the CACHENOUN file contains the most frequently used nouns. Finally, each file will be alphabetized. The initial sort--into categories--must be done by human hand, for the distinction between parts of speech is what we must "teach" the computer. The frequency sort will first be done by human intervention since initially the machine does not "know" how often different words are apt to be used. But, after the computer acquires some "experience", words will be shifted between NOUN file and CACHENOUN file, as certain nouns are seen to be used more frequently than other nouns (and the same is true for ADJ and CACHEADJ files, and so on). The alphabetization, however, can be done completely by machine, with neither "knowledge" nor "experience", and before any parsing starts.

There are many algorithms for sorting data alphabetically--three principal methods (insertion, exchange, selection), with at least three variations on each. The study of different sorts and the accompanying searches is taken up in a course in data structures. We will detail only one sort--an exchange sort known as the "bubble sort".

The bubble sort has the distinct advantage of being relatively easy to understand. It has the distinct disadvantage of being very slow. Despite this disadvantage, it is instructive for us to detail the construction and processing of the bubble sort.

Let us assume that we would like to sort six numbers: 3, -1, 8, -5, 0, 2, from the lowest to the highest. The principle behind the bubble sort is to look at two numbers at a time and switch them if the higher occurs first. Thus, our sort looks at 3 and -1 and switches them; so, out first move results in

-1, 3, 8, -5, 0, 2.

The sort then looks at 3 and 8. Since they are in the proper order, nothing is switched. Next come 8 and -5 which are switched to

-1, 3, -5, 8, 0, 2.

Next, 8 and 0 switch to

-1, 3, -5, 0, 8, 2.

Finally, 8 and 2 switch to

-1, 3, -5, 0, 2, 8

Note that we have made five comparisons, four of which resulted in switches. The result is that the largest number, 8, has been "bubbled" to the end of the set.

We now go through the process of comparison and exchange again, but we may ignore the last number since it is already in its proper place. Thus, we have only four comparisons, three switches, and a result of

-1, -5, 0, 2, 3, 8.

Now, 3 and 8 are in their proper places, so we need only three comparisons. Quite by accident, 0 and 2 are also in their proper places, but the computer does not know this and goes through the process as though they were not properly placed. After the third pass we get

-5, -1, 0, 2, 3, 8.

At this point we have the set completely ordered, but *the computer does not know*! It now wastes time going through two more passes, with *two* and *one* comparisons, respectively, and makes no exchanges. The set is now definitely sorted. Note that for 6 pieces of data we have 5 passes and one less comparison per increasing pass. A simple bubble sort routine is:

```
100         FOR PASS = 1 TO (N-1)        ! N elements to be sorted
110            FOR PLACE = 1 TO N-PASS ! places compared
120               IF A(PLACE)>A(PLACE+1) THEN GOSUB 500
130            NEXT PLACE
140         NEXT PASS
500      ! SUBROUTINE EXCHANGE
510         LET S = A(PLACE)            ! S stores a value
520         LET A(PLACE)=A(PLACE+1)
530         LET A(PLACE+1) = S
```

Lines 100-140 accomplish all of the necessary comparisons. The subroutine 500-540 exchanges values when the higher is to the left of the lower. The variable S is a necessary place-holder since we *lose the value* of A(PLACE) at line 520.

> Imagine a large bag of few peanuts and a small bag of many gumdrops. In order to switch bags, we cannot just throw the peanuts into the gum-drops. We need a third bag to hold the peanuts, put the gumdrops in the empty peanut bag, and finally put the stored peanuts into the gumdrops bag. The third bag is once again empty, but was necessary for the trans-fer. Thus, the need for our variable S.

To alphabetize, one only need change A to A$ and S to S$. The computer automatically sorts strings letter by letter in order of the ASCII or EBCDIC (used in most IBM machines) code of each letter.

A very important method of sorting and searching is with binary trees. Although it is outside of the scope of our text to produce algorithms for such sorts and searches, it is instructive for us to investigate trees and tree sorts and searches in a bit more detail than we have thus far.

A "binary tree" is a data structure in which the elements are connected at ancestral "nodes", each of which has a left descendant and a right descendant. The term "binary" is a result of each node having at most two descendants.

Here: John, Henry and Mary are located at ancestral nodes; Eric, Alfred, John and Helen are not. Henry and Mary are descendants of John; Eric and Alfred are descendants of Henry; John and Helen are descendants of Mary.

Once we know how Henry and Mary stand to John, we know how they stand to each other ("sibling"). Also, we know that the father of Eric's father is Eric's grandfather.

We have seen a simple binary tree in the organization of the number of A, B, C, D, and F grades in a class (Section 2.12), but let us look at a more interesting tree--a tree of Alphonse's statement "Marx sees a majority formed by people with commodiousness, expropriated in a purge".

The words of Alphonse's statement are randomly ordered as far as the computer is concerned (i.e., they do not follow the ASCII or EBCDIC sequence), although they are intensionally ordered with respect to English syntax. In order for the computer to find the word "with", it would have to check each word in order, starting from "Marx" until it reached "with"; which is the same linear process that it would go through for a random sequence of words. In this case, "with" is found on the eighth comparison. Of course, we could alphabetize the words and get:

a
a
by
commodiousness
expropriated
formed
in
majority
Marx
people
purge
sees
with

which would be better for some purposes, but worse for our search for the word "with"--that is , if we were using a linear search--for now "with" is found after 13 comparisons. The best sort would be an alphabetical tree sort:

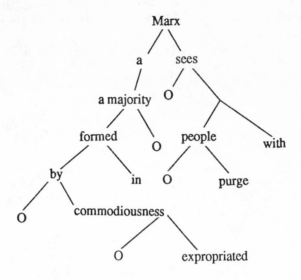

Although this tree seems very complicated to us, it is embarassingly easy for a computer to run through it. The algorithm of search, which is isomorphic to the algorithm of organization, is simply:

(1) Start at the top.
(2) If the given word (in this case "with") is (alphabetically) "greater than" the word at the top then go to the right node. If it is less than, then go to the left node.
(3) At the new node, repeat the process until
   A) you hit the word, or
   B) no branch remains (in which case the word is not in the list).

In this tree, "with" is found after 3 comparisons (i.e. a path of length 3). Notice that in our tree, "expropriated" needs the maximum path or the maximum number of comparisons (7) which is better than the random search (10), but worse than the alphabetical (4). But, most words need fewer comparisons in the tree sort than in either of the other two. Most importantly, a word not in the list can always be discovered in at most 8 comparisons in the tree, but needs 13 in the random and from 1 to 13 in the alphabetical.

Of course, the above tree is not the only one which can be formed; we could start with a different word, or we could tree diagram the alphabetical listing, such as

which is our worst case scenario. In fact, for a single word search such as we have above, the binary tree is at worst alphabetical and generally better than either random or alphabetical--unless, of course, one has a stroke of luck.

With this brief introduction to the problems and possible solutions of semi-semantic analyzers, we can begin to approach the forefront of automatic analyzers: complete semantic parsers.

Exercises
1. People organize their paperwork in various ways. Some let papers pile up on their desk, so that all information is naturally sorted by time of arrival (the first to arrive is at the bottom). Others create arbitrary categories and shove papers into

folders and the folders into a file cabinet. Discuss the advantages and disadvantages of each of these two sorting systems. Relate different searches to these sorts.
2. The following pieces of data are to be bubble sorted from lowest to highest.
   i) determine the order of elements after each pass.
   ii) determine how many passes will be made.
   iii) determine how many wasted passes are made.
       A) 8, -5, 13, 12, 7.
       B) 6, 6, 6, 6, 6, 6, 6, 6, 6.
       C) 4, -4, 4, -4, 4, -4.
       D) 10, 9, 8, 7, 6.
3. Sort the following (a) alphabetically, (b) in a binary tree. Determine the maximum path-length of the binary tree.
       A) To be or not to be - that is the question.
       B) Toot, toot, tootsie, goodbye.
       C) Two roads diverge in a yellow wood.

3.3 Semantic Categorizing

A semantic language analyzer constructs a structured set of terms, where the xxxyyy elements are distributed so as to be available as parts of an interactive conversation. It presupposes the existence of a syntactic (word related to word) structure and goes on to verify the relationship of word to meaning (=semantics).

    A category system is such a structure.

    When we indicated in Chapter I that logical functors differ not only among themselves but also relative to the terms that are their arguments, we were really suggesting that the categorial status of the arguments is not insignificant for a logical analysis of discourse.

    So, let us put together:

(1) a conversational program (i.e., an input-output sequence, where machine and "tutor" can question each other about the world), and
(2) a matrix of words, arranged in such a way that the interlocutors can discourse about the world (their "world of discourse").

Even though what we do here overcomes the limitation of the "semantic rings" approach by associating several of them around the same axis, there still are certain very rigid restrictions:

1. The input is typed. Therefore, the computer receives only the electronic representation of words, made up of representations of letters. It does not "understand", and it does not have the "unity" that perception confers on ordinary human experience. The tutor, on the other hand, perceives and contextualizes what he perceives.
2. At the outset, the computer memory contains only the ability to question. It has no other content; but acquires this latter in interaction with the tutor. The tutor, for his part, is already armed with many inter-related words and expressions.
3. During the early stages, there can be only a single tutor, using only a single categorial system. Otherwise, the data will be entered in a confused manner.

Once a word-stock of adequate size is endowed with one consistent structure, it can be used to acquire another type of structuring. In other words, once one knows English, Chinese can be learned, or vice versa. Both cannot be learned perfectly simultaneously.

The basic unit for our semantic parser is the set of 255 bytes, called a *record*. The record is divided into *fields* which we can use for nouns, verbs, adjectives and adverbs as follows:

word	freq	syntax	semantics	history	synonymy	
(1)	(2)	(3)	(4)	(5)	(6)	
0	25	28	38	68	99	225

(1) We allow 25 bytes to represent the word in question (any words that exceed these limits will be handled by a BIG-WORD subroutine). It is assumed that the plural of any word is "word + 's'". For irregular plurals, we set Byte 38 to 1 and put the plural form as last synonym.

(2) The frequency is increased by 1, until it reaches 9999, whereupon the counter is disabled and the word is assigned to the corresponding CACHE file (= a file of frequently used words).

(3) For the syntactic indicators, we allocate the Bytes between 29 and 38. Each of these can take one of the 10 values (1 to 9, and 0), for a total of 100.

(4) Byte 29 indicates the part of speech (1 = noun; 2 = verb; 3 = adjective; 4 = adverb; 5 = words commonly used as either adjective or adverb; 6 = pronouns; 7 = conjunctions; 8 = prepositions; 9 = interjections; 0 = unallocated).

(5) Byte 38 was allocated above to signalling irregular plurals.

The part of speech (Byte 29) determines which of the remaining bytes is used and for what.

For example, if Byte 29 is 1 (=noun), then Byte 30 can be set to 0 for an abstract noun (e.g. love) and to 1 for a concrete noun (e.g. book); and Byte 31 will be 0 for a noun that appears to be singular but actually is plural ("every" is a case that interests the logician in particular), and 1 for a noun that appears plural but acts as a singular (e.g. economics, measles).

If Byte 29 is 2 (=verb) then Byte 30 is set to 0 for a regular verb and to 1 for an irregular verb; Byte 31 then takes 1 for monadic, 2 for dyadic, 3 for triadic, and so on; Byte 32 is 1 for a prepositional verb (e.g. go through) and 0 for a non-prepositional verb; Byte 33 assigns the verb to its class (according to how the principal parts are formed); Byte 34 signals a subjunctive ("were" for the verb "to be").

As we will see below, the other parts of speech demand records that are much less complicated.

3.31 General Operation of a Semantic Categorizer

1. input a sentence
2. break it down into its components (words)

3. alphabetize them, keeping a record of their original order
4. do the check for syntactic correctness

A syntax checker will return the result
(a) anomalous expression (check it against a list of idioms and exceptions); or
(b) "no sentence"; ask the tutor to correct it or to type in a new sentence; or
(c) good syntax.

5. If we have "good syntax", then go on to the semantic categorizer, where we have filled in the major fields, somewhat along the lines of the Marx matrix.

(1)	(2)	(3)	(4)	(5)	(6)

0   38   68

Bytes 39 through 68 are distributed into three semantic subfields, each of which opens with a bit (39,49,59), each of which is set to 1 for a word having to do mainly with nature, to 2 for a word having to do mainly with something human, and to 3 for a word designating what is beyond both nature and the human.

Each of the three subfields is further subdivided into three trios of bits, representing the main classification, the main sub-classification, and the secondary classification. So, to take as an example the primary semantic subfield (Bytes 40 to 48):

Bytes 40, 43 and 46 take numbers corresponding to the ten questions (along the left side of the Marx matrix, and in specification 2 of Section 3.4 below).
Bytes 41, 44 and 47 take
    1   for the physical realm,
    2 for the biological realm,
    3 for the psychic (or 'personal') realm,
    4 for the cultural or spiritual realm.
Bytes 42, 45 and 48 go to

    1   if the word indicates what causes something to come
        to be;
    2   for what describes what something is composed of;
    3   if the form or shape is indicated;
    4   if the word says the object, end or purpose of some
        act.

The secondary and tertiary semantic subfields are structured in exactly the same way; and serve to define "cross-categorization" which is also what is generally meant by the "figurative" or "allusive" use of words.

(1)	(2)	(3)	(4)	(5)	(6)

0   68   99

The categories of history repeat the semantic fields and subfields. While the semantic categories define the status of the word itself and change little or not at all after their initial entry (primitive definition), those of the history change in accord with the words with which this word "associates"--thus specifying, for example, the professions, sciences, etc., where the word tends to occur in the vocabulary of this speaker.

(1)	(2)	(3)	(4)	(5)	(6)

0   99   225

Byte 99 can be used to count the total number of synonyms.
Bytes 101-125, 127-150, 152-175, 177-200, 202-225 will hold the synonyms (with the last being reserved for irregular plurals when such occur).
Bytes 100, 126, 151, 176, and 201 serve as counters for the occurrence of each synonym.

6. Formulate a response, using the words and terms that fall into "natural" correlation with the words and terms in the

sentence (using some variant of "semantic rings"). Essentially, the computer can be forced to "call up" or "construct" a three-dimensional matrix (along the lines of our Marx matrix) which can be made to vary according to the shifting correlations of syntactic constructs, semantic configurations, and cross-categorizations.

7. Get the next sentence and do the same.

3.4 Dictionaries and Lexicons

If the table look-up we mentioned above has to be done for every use of every word in the language, no computer could keep all this in storage (memory overload) or use it quickly enough to be practical (deteriorated response-time). What is more, if the syntactic look-up has to be "doubled" by one or more semantic look-ups, the whole process will be interminable.

We introduce the notion of a "poly-categorially structured dictionary" (PCSD), with the following specifications:

1. The major files (lists) of words are syntactically organized. For example: a file of words that are used exclusively as nouns (e.g. proper names); words that are just co-ordinating conjunctions ("and"); just adjectives; just adverbs; words used both as adjective and as adverb; and so on.

2. Within each file the records are arranged according to the ten questions (of the table at the end of Section 3.11):

   (1) is this a thing?
   (2) is it the main property of a thing?
   (3) is it a quantity? (how much, how many, etc.)
   (4) is it a relative term?
   (5) does it name place?
   (6) does it say when?
   (7) is it an activity?
   (8) is it something that is done to something else?
   (9) does it say how the parts of a thing are arranged?
   (10) does it describe something habitual or usual about the thing?

3. A small program that reads *freq* will regularly update the

files, putting the most frequently used *records* in the most accessible position (speeding up response-time), relegating seldom used terms to a file of *exceptions* (or *idioms*), and--most importantly--putting into an *immediate* (called *cache*) *memory* the words that are used so frequently that there is no point in counting them (there are 600 to 800 in English, according to some accounts, and most of them are what we called in Section 2 "syncategorematic" terms).

4. Thematic configurations of terms (our semantic rings, but now in the form of matrices) will result from the fact that (just as when one person talks with another person), once the clue to the "realm of discourse" has been detected, the correlations of categories, subcategories, etc. enable a quick orientation to the terms that are appropriate for a response.

In other words, when a live person talks with another live person he gives some clue as to what the subject matter is. If the originator says "Let's talk", the respondent will likely say "About what?". When the originator says "About logic", the respondent immediately "becomes sensitive" to the terms that are involved in this field. Similarly, if we type to the semantic parser "Let us talk about logic", a simple program allows it to ask "Are we going to talk about 'logic', 'us', 'talk', 'let', or 'about'?". (Note that the terms are arranged in order of likelihood: noun, pronoun, copula, preposition); just as the name "Marx" keyed in Alphonse in our earlier example.

5. Finally, in the initial stages of "person-categorizer interaction" the person will have to remain the same. This is to guarantee a certain consistency in the use of words, categories, grammar, etc.; i.e. to enable the semantic parser to "learn", meaning to achieve sets of matrices that make intelligible answers possible.

* * *

The PROLOG computer language we used in some simple examples above, got its name from PROgramming in LOGic. Here is part of a PROLOG version of a poly-categorially structured dictionary.

% SEMANTIC DIMENSION/VALUE STRUCTURE

dimension(language,1,meaning).
value(meaning,2,semantics).
value(range,1,infrahum).
value(range,3,suprahum).
value(ariscat,1,substance).
value(ariscat,3,quantity).
value(ariscat,5,action).
value(ariscat,7,where).
value(ariscat,9,habitus).
dimension(substance,1,katauto).
dimension(substance,3,katapantos).
dimension(quantity,1,measure).
dimension(relation,2,predicamental).
dimension(passion,1,emotion).
dimension(when,1,year).
dimension(dispositn,1,akimbo).
value(causal,2,formal).
value(causal,4,final).
dimension(suprahum,1,ariscat).

value(meaning,1,syntax).
dimension(semantics,1,range).
value(range,2,human).
dimension(infrahum,1,ariscat).
value(ariscat,2,quality).
value(ariscat,4,relation).
value(ariscat,6,passion).
value(ariscat,8,when).
value(ariscat,10,dispositn).
dimension(substance,2,katolou).
dimension(quality,1,essential).
dimension(relation,1,transcend).
dimension(action,1,causal).
dimension(where,1,global).
dimension(habitus,1,art).
value(causal,1,efficient).
value(causal,3,material).
dimension(human,1,ariscat).

% SYNTACTIC DIMENSION/VALUE STRUCTURE

dimension(syntax,1,term).
value(term,2,variant).
root(a,1,a).
root(am,1,be).
root(animal,1,animal).
root(are,1,be).
root(be,1,be).
root(being,1,be).
root(bird,1,bird).
root(bone,1,bone).
root(brown,1,brown).
var(animal,1,singular,animal).
var(be,1,infinite,be).
var(be,1,past-plural,were).
var(be,1,plural-present,are).
var(be,1,third-singular-present,is).

value(term,1, root).
value(term,3,word).
root(all,1,all).
root(an,1,an).
root(animals,1,animal).
root(ate,1,eat).
root(been,1,be).
root(big,1,big).
root(birds,1,bird).
root(bones,1,bone).
var(animal,1,plural,animals).
var(be,1,1st-singular-present,am).
var(be,1,past-participle,been).
var(be,1,past-singular,was).
var(be,1,present-participle,being).
var(bird,1,plural,birds).

var(bird,1,singular,bird).
var(bone,1,plural,bones).
var(bone,1,singular,bone).
var(cat,1,plural,cats).
var(cat,1,singular,cat).
var(cow,1,plural,cows).
var(cow,1,singular,cow).
var(do,1,infinitive,do).
var(do,1,past,did).
var(do,1,past-participle,done).
var(do,1,present-participle,doing).
var(do,1,3rd-singular-present,does).
var(dog,1,plural,dogs).
var(dog,1,singular,dog).
var(eat,1,infinitive,eat).
var(eat,1,past,ate).
var(eat,1,past-participle,eaten).
var(eat,1,present-participle,eating).
var(eat,1,third-singular-present,eats).
word(a,1,part_of_speech,determiner).
word(all,1,part_of_speech,determiner).
word(an,1,part_of_speech,determiner).
word(animal,1,collectivity,non-coll.)
word(animal,1,commonality,comm.).
word(animal,1,concreteness,concrete).
word(animal,1,part_of_speech,noun).
word(be,1,part_of_speech,copula).
word(big,1,part_of_speech,adjective).
word(bird,1,collectivity,non-coll.).
word(bird,1,commonality,comm.).
word(bird,1,concreteness,concrete).
word(bird,1,part_of_speech,noun).
word(bone,1,collectivity,non-coll.).
word(bone,1,commonality,comm.).
word(bone,1,concreteness,concrete).
word(bone,1,part_of_speech,noun).
word(brown,1,part_of_speech,adj.).
word(cat,1,collectivity,non-coll.).
word(cat,1,commonality,common).
word(cat,1,concreteness,concrete).
word(cat,1,part_of_speech,noun).
word(cow,1,collectivity,non-coll.).
word(cow,1,commonality,comm.).
word(cow,1,concreteness,concrete).
word(cow,1,part_of_speech,noun).
word(do,1,part_of_speech,verb).
word(do,1,type,do).
root(plane,1,plane).
var(plane,1,singular,plane).
word(plane,1,part_of_speech,noun).
root(planes,1,plane).
var(plane,1,plural,planes).
word(plane,1,commonality,comm.).
word(plane,1,concreteness,concrete).
word(plane,1,collectivity,non-coll.).
root(machine,1,machine).
var(machine,1,singular,machine).
word(machine,1,part_of_spch,noun).
root(machines,1,machine).
var(machine,1,plural,machines).
word(machine,1,common.,comm.).
word(machine,1,concreten.,concrete).
word(machine,1,collect.,non-coll.).
root(fighter,1,fighter).
var(fighter,1,singular,fighter).
word(fighter,1,part_of_speech,noun).
root(fighters,1,fighter).
var(fighter,1,plural,fighters).
word(fighter,1,commonality,comm.).
word(fighter,1,concreteness,concrete).
word(fighter,1,collectivity,non-coll.).
root(i,1,i).
var(i,1,singular,i).
word(i,1,part_of_speech,pronoun).
root(we,1,i).
var(i,1,plural,we).
word(i,1,person,first).
word(i,1,case,subjective).
word(i,1,question,no).

Note that 'word', 'root', and 'var' give us the flexibility needed to specify and distinguish not only parts of speech but also irregular usages. Also, because of PROLOG's "backtracking" characteristic (a form of "recursion"), no complicated sorting routines are needed.

3.5 Toward an Artificial Intelligence

While it is beyond the scope of this introduction to computing, logic and mathematics to construct a complete language analyzer, we should be clear on precisely how this can be done.

Instead of continuing with the Marxian vocabulary--which is not necessarily everyone's cup of tea--let us pursue our analysis in our chosen field, logic.

*   *   *

Alphonse: "Good morning, Gaston. Why did you activate the A.C. and load the Operating System?"
Gaston: "I have a problem."
(Alphonse registers an anomaly when "have" is followed by an abstract term, rather than a concrete noun. It consults IDIOM.DAT to find "have a problem with" and, correspondingly, can formulate the question...)
Alphonse: "You have a problem with what?"
Gaston: "I have a problem with logic."
(Note: the more natural answer "With logic." could easily be accomodated by ENCLITIC.BAS, a program that assumes as present anything in the question preceding the first word that occurs in the answer. However, since we are currently using the convention of typing in complete sentences, enclitic sentences are not a problem.)
Alphonse: (1) "Logic deals with ...., ..., and ...." (picking up on correlative words that have co-occurred with "logic" in previous conversations),

OR

(2) "I have no files with words about logic. Tell me about it."

In the case of (2), obviously, Alphonse will engage in a thorough questioning of Gaston in order to fill in as many fields as possible in the record of "logic", thus inserting it into Alphonse's active vocabulary, but also establishing its correlation with terms that are already there.

What about (1)? In other words, how when Gaston invokes "logic" is Alphonse able to come up with correlative words like "inference", "deduction", and "implication"?

On Gaston's mention of "logic", Alphonse then

1. looks for "logic" in CACHENOUN.DAT, one of a series of files that contain the words most frequently used by Alphonse (others include CACHEVERB.DAT, CACHEADJADV.DAT, CACHEABS.DAT (this last contains all conjunctions, prepositions, pronouns); failing which it
2. looks for "logic" in ALLWORDS.DAT which contains, in alphabetical order, all Alphonse's words (except those which are already in a CACHE).

Wherever "logic" is found--and we assumed in 1. that it is found--the semantic fields, history and synonymy contain indices that allow the appropriate file(s) to be ordered according to the relative frequency of correlatives of "logic".

"Logic" turns out to be a *noun*, which is
abstract, belonging to the semantic level of the
human, more precisely the human
higher operations, which are
cognitive, and
relational, etc., etc., having been previously used as
science,
related to mathematics, and in
philosophy, etc., etc., with
"*science on thought*" and "*science on form*" as synonyms.

There results, then, a number which is used to sift ("bubble") to the top of all files (NOUNS.DAT, VERBS.DAT, ALLWORDS.DAT, etc., etc) all the words with the same number(s), beginning with the most frequently encountered. The remainder of the files stay in alphabetical order. Here is an incomplete example:

logic

elements	operations	properties
term(s)	induction	true
word(s)	inference	valid
subject	deduction	false
predicate	definition	correct
copula	division	soundness
symbol	association	bound
variable	truth-table	completeness

So, when Gaston says "I have a problem with logic", Alphonse finds at the top of the noun file the nouns that have correlated with logic in the past. A similar reordering is done for verbs, adjectives, adverbs--setting up, in effect, an immediate stock of the words that are the most useful for talking about logic.

Instead of mere "semantic rings", we now have "semantic arrays" which make possible a richer conversation on a wider range of topics.

* * *

Beyond the conversational use of our analyzer lies the possibility of using it on a piece of written text. Here is Aristotle's original description of syllogistics:

> Whenever three terms are so related to one another that the last is contained in the middle as in a whole, and the middle is either contained in, or excluded from, the first as in or from a whole, the extremes must be related by a perfect syllogism. I call that term middle which is itself contained in another and contains another in itself: in position also this comes in the middle. By extremes I mean both that term which is itself contained in another and that in which another is contained. If A is predicated of all B, and B of all C, A must be predicated of all C: we have already explained what we mean by 'predicated of all'. (25b32-26a1)

We proceed as follows:

1. Find the first terminal punctuation ("...perfect syllogism. I

call...") and initial capital ("Whenever...").
2. Find the first verb ("are") in CACHEVERB.DAT.

> Rule 1 for "to be": see if a participle precedes or follows it. (No)
> Rule 2 for "to be": is what follows it of the same syntactic type as
> what precedes it (ignoring articles, prepositions, etc.)? (No)

3. Return to Rule 1, checking "are-2" and "are+2"
   ("are...related" = a   valid copula), found in ALLVERB.DAT.
4. [subject = "Whenever three terms"; copula = "are...related";
   predicate = "to one another..."]
5. Do the Subject: find the substantive ("...terms...") in
   CACHENOUN.DAT or ALLNOUN.DAT.
6. Finish the Subject: check PROPNAME1.DAT and
   PROPNAME2.DAT for "Whenever" (because it begins with
   a capital letter) (No) Check CACHEADJ.DAT for
   "Whenever" (Yes, and we are sent to ALLCONJ.DAT,
   where we learn that...)
7. "Whenever" is a subordinating conjunction that introduces a
   dependent clause; so
8. Interrupt processing of this clause in order to locate and
   process the independent clause.
9. "Whenever" sends us to find the comma (or semi-colon) that
   separates dependent from independent clause ("..., and ..."):

Find the verb ("... is ...")

> Note: *we*, of course, know that it would be more efficient for Alphonse
> to see the "and" that follows the first comma and then the "or"
> following the second, and to "realize" that these are "compound"
> conditions. However, Alphonse does not "realize" or "understand"
> anything. All there is is a sequential check and an algorithm, according
> to which the sequential check is applied first to find the copula, then to
> process the subject, and then to do the predicate. In any case, the third
> comma ("...from, the...) would defeat a simple algorithm for detecting
> such cases of compound sentences.

> Rule 1 for "to be": ... (No)
> Rule 2 for "to be": ... (No)

11. Return to Rule 1, checking "is-2" and "is+2" ("is...contained in" is a valid copula)

(we might want to learn from this second occurrence of a split copula that it would be useful to have a rule for handling them)

12. (Subject = "and the middle"; Copula = "is ... contained in"; Predicate = "0")
13. Since there is no predicate, this cannot be the independent clause (required by "Whenever"). Interrupt processing this clause, and
14. Find the comma (since 9 above is unsatisfied)(="...contained in, or...")
15. Find the verb ("...excluded...")
16. But, since "exclude" is dyadic and there is nether subject nor predicate, this is not the independent clause. Interrupt this and
17. Find the comma (since 9 above is still unsatisfied) (= "from, the...")
18. Find the verb (No)
19. Find the comma ("...whole, the...")
20. Find the verb ("must"). And, since "must" requires it,
21. Find the needed complement ("must be related")
22. (Subject = "the extremes"; Copula = "must be related"; Predicate = "by a perfect syllogism")
23. Do the subject: find the substantive ("extremes")
24. Do the Predicate: since "related" is a modal verb and "by" introduces a phrase of manner, the independent clause (required by "Whenever") has been successfully located and analyzed.
25. At this point, Alphonse has "Whenever...whole," as antecedent and "the extremes...syllogism." (a valid dependent clause) as consequent, and will issue a "good-sentence" conclusion.

We could, of course, design algorithms for handling complex and compound sentences, taking the clues from the "and(s)", "or(s)" and "either(s)" that were left unanalyzed above. However, it is quicker to leave these elements to the semantic parser, where they will be checked only if some nonsense occurs.

26. Find the thematic area, by examining first all verbs (related, contained, excluded, call, comes, mean, predicated, explain) (nothing special here), then all nouns (terms, middle, extremes, syllogism, A, B, C) (clearly "logic" is the domain).

At this point, Alphonse needs an interlocutor or a specific task to execute. For instance, it could be used to discover the dominant trend in the words, to catalogue the imagery used, etc.

Exercises:
1. Do the same sort of analysis with one of the remaining sentences, from Aristotle's text.
2. Show how subroutines could be used to reduce repetition in the above analysis.

* * *

Here are the file-structures in probable order of access:
   Since we begin with the copula, we check
   CACHEVERB.DAT, then
   ALLVERB.DAT, pointing to
   MONADREG.DAT, DYADREG.DAT, etc. for regular verbs, and
MONIRREG.DAT, DYIRREG.DAT, etc. for irregular verbs. The
VERB-RECORD contains the same information as the
   NOUN-RECORD, plus a listing of the "principal parts" (e.g. go, went, gone) and any irregular forms (especially the subjunctive, if there is one).
   We deal with participles (present and past), gerundives and infinitives as mainly adjectival.

   To do substantives (in the Subject and/or the Predicate):
   When the first letter is a capital, check
   PROPNAME1.DAT for Christian (or first) names, and
   PROPNAME2.DAT for family names, ELSE check
   CACHENOUN.DAT (= most frequently used nouns), then
   ALLNOUN.DAT, which contains pointers to the structured noun files.

To do other words (in the "rest of Subject" and "rest of Predicate"): Check

CACHEABS.DAT, which contains pointers to:

ALLPRONS.DAT(personal: I(me), you, he(him), she(her), it, we(us), they(them); possessive: my(mine), your(s), his, her, hers, its, our, ours, their, theirs; reflexive: myself, yourself, himself, herself, itself, ourselves, yourselves; relative: who, whom, which, that, whose; interrogative: who, whom, which, what, whose; demonstrative: this, that, these, those; indefinite, or to

ALLCONJS.DAT (subordinate or co-ordinate; sentential or nominal) or to

ALLPREPS.DAT, where the prepositions are ordered from multiple to simple; then check

CACHEADJADV.DAT, pointing to

ALLADJADV.DAT, containing adjectives (ending in -al, -ive, -ful, -ing, -ed, -er, -est) and the adverbs derived from them (ending in -ly, -wise, -ways); and

ALLRSTADV.DAT, with the rest of the adverbs (state of mind (willingly), manner (broadly), evaluative (clumsily), resultant (correctly), then check

ALLWORDS.DAT, containing pointers to files with all the other words. This is an alphabetical master list that is checked as last resort, when all the more likely checks fail.

ALLEXCLS.DAT, consultation of which is tripped by a terminal "!" or double commas. Exclamations are: only initial, only median, only terminal, only alone; and one-word, two-word, three-word, etc.

MISSPELL.BAS consults MISSPELL.DAT for common misspellings, before declaring a word to be out of bounds.

IDIOM.BAS looks at IDIOM.DAT when there is an anomalous expression.

*   *   *

Here are some sample record-structures:

Full records for nouns, adjectives (especially if noun-related), verbs (with above exceptions) and adverbs (especially when adjectivally derived).

174

PROPNAME1.DAT and PROPNAME2.DAT need only be marked as nouns. (the main clue is the initial capital letter)

ALLPREPS.DAT needs only syntactic information.

ALLCONJS.DAT contains only the functional information.

ALLPRONS.DAT must be carefully structured because pronouns play many propositional roles.

MISSPELL.DAT = incorrect(s) → correct(s)

IDIOM.DAT = in alphabetical order? according to first word? or according to most significant word?

ALLVERBS.DAT, ALLNOUNS.DAT, CACHEABS.DAT, ALLWORDS.DAT will all quickly accumulate so many words that they will have to be declared pointer files from the beginning.

Appendix 1
Matrix of Valid Syllogistic Modes

In what follows, we indicate only one error for each configuration. In order of logical importance, the errors are: faulty middle term (BAD MID); conclusion is positive or universal when a premiss is negative or particular (P<>D, S<>D); error of distribution (WEAKER); two particular premisses (2 PARTS); two negative premisses (2 NEGS). Of course, sometimes the opposite order is easier to notice. In other words, one sees immediately two negatives and/or two particulars.

	M-P S-M S-P	P-M S-M S-P	M-P M-S S-P	P-M M-S S-P
AAA	BARBARA	BAD MID	S<>D	S<>D
AAE	P<>D	BAD MID	SP<>D	S<>D
AAI	BARBARI	BAD MID	DARAPTI	BRAMANTIP
AAO	P<>D	BAD MID	P<>D	pos->neg
AEA	WEAKER	WEAKER	WEAKER	WEAKER
AEE	P<>D	CAMESTRES	P<>D	CAMENES
AEI	WEAKER	WEAKER	WEAKER	WEAKER
AEO	P<>D	CAMESTROP	P<>D	CAMENOP
AIA	WEAKER	WEAKER	WEAKER	WEAKER
AIE	WEAKER	WEAKER	WEAKER	WEAKER
AII	DARII	BAD MID	DATISI	BAD MID
AIO	P<>D	BAD MID	P<>D	BAD MID
AOA	WEAKER	WEAKER	WEAKER	WEAKER
AOE	WEAKER	WEAKER	WEAKER	WEAKER
AOI	WEAKER	WEAKER	WEAKER	WEAKER
AOO	P<>D	BAROCO	P<>D	BAD MID
EAA	WEAKER	WEAKER	WEAKER	WEAKER
EAE	CELARENT	CESARE	S<>D	S<>D

EAI	WEAKER	WEAKER	WEAKER	WEAKER
EAO	<u>CELARONT</u>	<u>CESARO</u>	<u>FELAPTON</u>	<u>FESAPO</u>
EEA	2 NEGS	2 NEGS	2 NEGS	2 NEGS
EEE	2 NEGS	2 NEGS	2 NEGS	2 NEGS
EEI	2 NEGS	2 NEGS	2 NEGS	2 NEGS
EEO	2 NEGS	2 NEGS	2 NEGS	2 NEGS
EIA	WEAKER	WEAKER	WEAKER	WEAKER
EIE	WEAKER	WEAKER	WEAKER	WEAKER
EII	WEAKER	WEAKER	WEAKER	WEAKER
EIO	<u>FERIO</u>	<u>FESTINO</u>	<u>FERISON</u>	<u>FRESISON</u>
EOA	2 NEGS	2 NEGS	2 NEGS	2 NEGS
EOE	2 NEGS	2 NEGS	2 NEGS	2 NEGS
EOI	2 NEGS	2 NEGS	2 NEGS	2 NEGS
EOO	2 NEGS	2 NEGS	2 NEGS	2 NEGS
IAA	WEAKER	WEAKER	WEAKER	WEAKER
IAE	WEAKER	WEAKER	WEAKER	WEAKER
IAI	BAD MID	BAD MID	<u>DISAMIS</u>	<u>DIMARIS</u>
IAO	BAD MID	BAD MID	P<>D	P<>D
IEA	WEAKER	WEAKER	WEAKER	WEAKER
IEE	WEAKER	WEAKER	WEAKER	WEAKER
IEI	WEAKER	WEAKER	WEAKER	WEAKER
IEO	P<>D	S<>D	P<>D	P<>D
IIA	2 PARTS	2 PARTS	2 PARTS	2 PARTS
IIE	2 PARTS	2 PARTS	2 PARTS	2 PARTS
III	2 PARTS	2 PARTS	2 PARTS	2 PARTS
IIO	2 PARTS	2 PARTS	2 PARTS	2 PARTS
IOA	2 PARTS	2 PARTS	2 PARTS	2 PARTS
IOE	2 PARTS	2 PARTS	2 PARTS	2 PARTS
IOI	2 PARTS	2 PARTS	2 PARTS	2 PARTS
IOO	2 PARTS	2 PARTS	2 PARTS	2 PARTS
OAA	WEAKER	WEAKER	WEAKER	WEAKER
OAE	WEAKER	WEAKER	WEAKER	WEAKER
OAI	WEAKER	WEAKER	WEAKER	WEAKER
OAO	BAD MID	P<>D	<u>BOCARDO</u>	P<>D
OEA	2 NEGS	2 NEGS	2 NEGS	2 NEGS
OEE	2 NEGS	2 NEGS	2 NEGS	2 NEGS
OEI	2 NEGS	2 NEGS	2 NEGS	2 NEGS
OEO	2 NEGS	2 NEGS	2 NEGS	2 NEGS
OIA	2 PARTS	2 PARTS	2 PARTS	2 PARTS
OIE	2 PARTS	2 PARTS	2 PARTS	2 PARTS

OII	2 PARTS	2 PARTS	2 PARTS	2 PARTS
OIO	2 PARTS	2 PARTS	2 PARTS	2 PARTS
OOA	2 PARTS	2 PARTS	2 PARTS	2 PARTS
OOE	2 PARTS	2 PARTS	2 PARTS	2 PARTS
OOI	2 PARTS	2 PARTS	2 PARTS	2 PARTS
OOO	2 PARTS	2 PARTS	2 PARTS	2 PARTS

Appendix 2

Axioms and Laws of Propositional Calculus

In what follows the { } do not indicate a set but are just the third type of enclosure, after ( ) and [ ].

$\{[(p{\to}q){\bullet}(r{\to}s)]{\bullet}(p{\vee}r)\}{\to}(q{\vee}s)$	CKKCpqCrsAprAqs
$[(p{\to}q){\bullet}p]{\to}q$	CKCpqpq
$(p{\to}q){\leftrightarrow}({\sim}q{\to}{\sim}p)$	ECpqCNqNp
$(p{\to}q){\leftrightarrow}({\sim}p{\vee}q)$	ECpqANpq
$[(p{\to}q){\bullet}{\sim}q]{\to}{\sim}p$	CKCpqNqNp
$[(p{\bullet}q){\bullet}r]{\leftrightarrow}[p{\bullet}(q{\bullet}r)]$	EKKpqrKpKqr
$(p{\leftrightarrow}q){\leftrightarrow}[(p{\to}q){\bullet}(q{\to}p)]$	EEpqKCpqCqp
$(p{\leftrightarrow}q){\leftrightarrow}[(p{\bullet}q){\vee}({\sim}p{\bullet}{\sim}q)]$	EEpqAKpqKNpNq
$[(p{\vee}q){\bullet}{\sim}p]{\to}q$	CKApqNpq
$[(p{\vee}q){\vee}r]{\leftrightarrow}[p{\vee}(q{\vee}r)]$	EAApqrApAqr
${\sim}(p{\to}q){\leftrightarrow}(p{\bullet}{\sim}q)$	ENCpqKpNq
${\sim}(p{\bullet}q){\leftrightarrow}({\sim}p{\vee}{\sim}q)$	EKpqANpNq
${\sim}(p{\vee}q){\leftrightarrow}({\sim}p{\bullet}{\sim}q)$	ENApqKNpNq
${\sim}{\sim}{\sim}p{\leftrightarrow}{\sim}p$	ENNNpNp
${\sim}{\sim}p{\leftrightarrow}p$	ENNpp
${\sim}(p{\bullet}{\sim}p)$	NKpNp
$({\sim}p{\to}p){\to}p$	CCNppp
$({\sim}p{\to}p){\to}q$	CCNppq
$(p{\to}{\sim}p){\to}{\sim}p$	CCpNpNp
$[(p{\to}q){\bullet}(p{\to}{\sim}q)]{\to}{\sim}p$	CKCpqCpNqNp
$[(p{\to}q){\bullet}(p{\to}r)]{\to}[p{\to}(q{\bullet}r)]$	CKCpqCprCpKqr
$[(p{\to}q){\bullet}(q{\to}r)]{\to}(p{\to}r)$	CKCpqCqrCpr
$\{[(p{\to}q){\bullet}(q{\to}r)]{\bullet}(r{\to}s)\}{\to}(p{\to}s)$	CKKCpqCqrCrsCps
$[(p{\to}q){\bullet}(r{\to}s)]{\to}[(p{\bullet}r){\to}(q{\bullet}s)]$	CKCpqCrsCKprKqs
$[(p{\to}r){\bullet}({\sim}p{\to}r)]{\to}r$	CKCprCNprr
$[(p{\to}r){\bullet}(q{\to}r)]{\to}[(p{\vee}q){\to}r]$	CKCprCqrCApqr
$(p{\bullet}{\sim}p){\to}{\sim}p$	CKpNpNp
$[p{\bullet}(q{\vee}r)]{\leftrightarrow}[(p{\bullet}q){\vee}(p{\bullet}r)]$	EKpAqrAKpqKpr

179

p→(pVq)	CpApq
p→(q→p)	CpCqp
[p→(q→r)]↔[q→(p→r)]	ECpCqrCqCpr
{[p→(q•r)]•(~q•~r)}→~p	CKCpKqrKNqNrNp
[p↔(q↔r)]↔[(p↔q)↔r]	EEpEqrEEpqr
[pV(pVq)]↔(pVq)	EApApqApq
[pV(qVr)]↔[(pVq)Vr]	EApAqrAApqr
(p•q)→p	CKpqp
[(p•q)→r]↔[p→(q→r)]	ECKpqrCpCqr
(p•q)↔(q•p)	EKpqKqp
[p•(q•r)]↔[(p•q)•r]	EKpKqrKKpqr
p→[(p→q)→q]	CpCCpqq
(p↔~q)↔(q↔~p)	EEpNqEqNp
(p↔q)↔(~p↔~q)	EEpqENpNq
(p↔q)↔(q↔p)	EEpqEqp
p↔p	Epp
(p↔q)→(p→q)	CEpqCpq
(p↔q)↔(~q↔~p)	EEpqENqNp
[pV(q•r)]↔[(pVq)•(pVr)]	EApKqrKApqApr
pV~p	ApNp
(pVp)↔p	EAppp
[(pVq)•~p]→q	CKApqNpq
[(pVq)•~q]→p	CKApqNqp
(pVq)↔~(~p•~q)	EApqNKNpNq
(pVq)↔(~p→q)	EApqCNpq
(pVq)↔(~p|~q)	EApqDNpNq
(pVq)↔(qVp)	EApqAqp
(p•p)↔p	EKppp
(p•q)→q	CKpqq
[(p•q)→r]↔[(~r•q)→~p]	ECKpqrCKNrqNp
[(p•q)→r]↔[(p•~r)→~q]	ECKpqrCKpNrNq
[(p•q)→r]↔[q→(p→r)]	ECKpqrCqCpr
(p•q)→(p→q)	CKpqCpq
(p•q)↔~(~pV~q)	EKpqNANpNq
(p•q)↔~(p→~q)	EKpqNCpNq
(p•q)↔~(p|q)	EKpqNDpq
(p|p)↔~p	EDppNp
(p|q)↔~(p•q)	EDpqNKpq
(p|q)↔(~pV~q)	EDpqANpNq
(p|q)↔(p→~q)	EDpqCpNq

[(p|q)•p]→~q

[(q→p)•~(r•~p)]→[(qVr)→p]

(q→r)→[(p→q)→(p→r)]

q→[(p|q)→~p]

CKDpqpNq

CKCqpNKrNpCAqrp

CCqrCCpqCpr

CqCDpqNp

The right column in our list of formulae uses the Lukasiewicz or "Polish" notation, that was introduced above. The "reverse Polish notation" is used in some computers (e.g. Hewlett-Packard). One strength of Polish notation is that all its symbols are found on the ordinary typing keyboard. Another strength is that it dispenses with brackets, points, or other typographical conventions. For example, once we know that N is the monadic functor for negation and C, A and K are the dyadic functors for implication, alternation and conjunction, respectively, and that functors are written from left to right in decreasing order of strength, then we can see how

CKCpqCqrCpr

is equivalent to

[(p→q)•(q→r)]→(p→r)

which is the rule of syllogistics:

We parse it as follows:

1. The leftmost functor, C requires two arguments.
2. Can K and/or C fill the bill? No, since each of them is dyadic and requires two arguments.
3. Cpq is the first satisfiable formula and becomes the first (or left) argument of K.
4. Cqr is the second satisfiable formula and becomes the second (or right) argument of K.
5. This conjunction of implications, KCpqCqr, becomes the first (or left) argument of the first C.
6. The last satisfiable formula, Cpr, can be nothing but the second argument of the first C, and the process is complete.

With some practice, one finds the Polish notation easier to

use than any other; and truth-table verification is even easier, as follows:

Tautology 5 that we discussed above

$(\sim pVq) \leftrightarrow (p \rightarrow q)$

comes out in Polish notation as

EANpqCpq

which can be verified as follows (using T for true and F for false):

EA	Npq	Cpq
T T	F T T	T T T
T F	F T F	F T F
T T	T F T	T F T
T T	T F F	T F F

We start by filling in the p and q columns consistently and work from right to left:

The C column is Cpq.

The N column is Np.

The A column is ANpq.

The E column is EANpqCpq, comparing the A and C columns, and resulting in a tautology. QED

A very uncomplicated truth-table verification program can be written for axioms and rules in Polish notation, somewhat as follows:

1. Read the last character.
   a. If it is a capital letter, quit, as every functor has to have at least one argument.
2. Read the second-last letter.
   a. If it is a capital letter, it is the functor of the last letter and can only be N because all the others require two arguments.
   b. If it is not a capital letter, it and the last letter are the two arguments of the third-last letter, which must be a capital.

3. Read the third-last letter and apply the relevant truth-table, resulting in a "T" or "F" for the functor.
4. Continue reading backwards and satisfying each functor by assigning to it the appropriate truth value.

Appendix 3

Solutions to some Exercises

0.8   Algorithms (p. 17)

Answer to *Exercise:*

Assuming the shoelaces are on the shoes,

```
100      FOR J = 1 TO 2                       ! a shoe at a time
110          GRASP LEFT LACE IN LEFT HAND
120          GRASP RIGHT LACE IN RIGHT HAND
130          COIL RIGHT LACE ONCE AROUND LEFT LACE
140          SWITCH HANDS AND TIGHTEN
150          FORM LOOP ON LEFT LACE WITH LEFT HAND
160          FORM LOOP ON RIGHT LACE WITH RIGHT HAND
170          COIL THE LOOPS ONCE, SWITCHING HANDS
180          TIGHTEN
190      NEXT J
```

1.2   Propositional Logic (p. 53)

Answers to *Exercises*

a)   Dogs bark or dogs do not bark.
b)   It is not the case that cats meow and do not meow.
c)   If dogs bark then, cats meow and frogs croak.
d)   Dogs bark or cats meow, if and only if cats meow or dogs bark.
e)   Not a proposition.
f)   If dogs bark then, if cats meow then dogs do not bark.
g)   If dogs bark and cats meow then dogs bark or cats meow.
h)   Not a proposition.

1.3 Quantifiers (p. 56)

Answers to *Exercises*

Note: in some cases the order of binding makes no difference in ordinary language, e.g. 1.G), or 2. A) and 2. D) or 2. C) and 2. F).

1. A) Everybody is mortal.
   B) There is at least one man.
   C) It is not true that everyone is mortal.
   D) There is someone such that everyone - if that first one is mortal - is human.
   E) For everyone there is someone such that if that one is mortal, they all are human.
   F) There is a mortal who is not human.
   G) There is someone such that for someone else, that the first one is mortal, implies that the second is mortal.
   H) There is someone such that for everyone, that the first one is mortal implies that that one is mortal or they all are human.

2. A) Everybody loves everybody.
   B) Somebody loves everybody.
   C) Somebody loves somebody.
   D) Everybody is loved by everybody.
   E) Everybody is loved by somebody or other.
   F) Somebody is loved by somebody.

2.12 Classification and Division (p. 76)

Answers to *Exercises*

1. Under short-haired and long-haired we have show dogs, which is not a function like hunting or work, so that "show dogs" involves
   a) changes of principle and

b) an overlap. The classifications are not exhaustive because, e.g. non-showable specimens bred for fighting are not covered.
2. Under domestic wines, cheap and expensive are mutually exclusive and exhaustive but (alas)
    a) vague categories; "in plastic jugs" marks
    b) a change of principle and is
    c) not exclusive (of cheap wines presumably). Sparkling wines are distinguished from still, not by color, so that imported wines involve
    d) a change of principle and
    e) overlap. Also claret is omitted so that the classification is
    f) not exhaustive. Note that it is inelegant but not incorrect to subdivide domestic and imported wines differently.
3. What is served in fast food places does not constitute a food group so that
    a) we have a change of principle and
    b) overlap (with meat and cereal groups presumably). Since dairy products are missing, the classification is
    c) incomplete.
4. Artist is not a nationality and might not exclude French and Germans, hence though the sentence is possible in bizarre circumstances, it is odd.
5. a) Dogs do not exclude dachshunds, nor
    b) do cats in the broad sense exclude panthers.
6. The sentence presents as equals the thing eaten and mode of eating.

2.2 Propositions (p. 82)

Answers to *Exercises* 2

1. All who smoke cigars during the symphony are boors.
2. All men in love are happy.
3. Some professors are pedantic.
4. Some students are not lazy. (Occasionally "All ... are not ..."

is used to mean "No ... is ...")
5. Some children do not pass up pizza.
6. No mothers abandon their children.
7. Some reptiles are water-dwellers.
8. No cows are carnivores.
9. All who understand the contrapositive are logicians.
10. Some professors are funny.

2. 31 Immediate Inference (p. 91)

Answers to *Exercises*

1. a. All who chase cats are bad dogs. All bad dogs chase cats.
   b. All S is P. All P is S.
   c. No relation
   d. Undetermined
   e. Undetermined
2. a. No cat swims. Some cats do not swim.
   b. No S is P. Some S is not P.
   c. Subalternation
   d. True
   e. Undetermined
3. a. All cats are tunafish eaters. No cat is other than a tunafish eater.
   b. All S is P. No S is non-P.
   c. Obversion
   d. True
   e. False
4. a. All cats go crazy with catnip. Some cats go crazy with catnip.
   b. All S are P. Some S are P.
   c. Subalternation.
   d. True
   e. Undetermined.
5. a. All cats wear bells. Some cats do not wear bells.
   b. All S are P. Some S are not P.
   c. Contradiction.
   d. False

e. True
6. a. Some cats are not striped. Some cats are striped.
   b. Some S are not P. Some S are P.
   c. Subcontraries.
   d. Undetermined.
   e. True.
7. a. Some cats are not selfish. All cats are selfish.
   b. Some S are not P. All S are P.
   c. Contradictions.
   d. False
   e. True
8. a. No tabby cats are cute. All tabby cats are cute.
   b. No S is P. All S is P.
   c. Contraries
   d. False
   e. Undetermined
9. a. All who like cats are women. Some women like cats.
   b. All S are P. Some P are S.
   c. Conversion
   d. True
   e. Undetermined
10. a. All who like Manx cats are eccentrics. All
       non-eccentrics are non-likers of Manx cats.
    b. All S is P. All non-P is non-S.
    c. Contrapositive
    d. True
    e. False
11. a. All Persian cats are non-purrers. No Persian cat
       purrs.
    b. All S is non-P. No S is P.
    c. Obversion
    d. True
    e. False
12. a. No Abyssinian cat is long-haired.
       Some non-long-haired (cats) are not non-Abyssinians.
    b. No S is P. Some non-P is not non-S.
    c. Contrapositive
    d. True
    e. Undetermined

1. No philosopher is crazy.
2. Some philosopher is crazy.
3. Some philosopher is not crazy.
4. Some crazy people are philosophers.
5. Some crazy people are not philosophers.
6. All philosophers are other than crazy.
7. No philosopher fails to be crazy.
8. All who are other than philosophers are other than crazy.
9. All who are other than crazy are other than philosophers.
10. No crazy people are philosophers.

2.311 Modal Logic (p. 95)

Answers to *Exercises*

1. It is impossible for Reagan and Gorbachev to reach an arms agreement.
2. Necessity and impossibility are contraries, so there is no contrary of a statement in the possible mode.
3. It is possible for some logician to commit a fallacy sometime.

2.32 The Assertoric Syllogism (p. 103)

Answers to *Exercises*

1. a. All philosophers drink hemlock.
      <u>All who drink hemlock irritate Athenians</u>.
      So All who irritate Athenians are philosophers.
   b. All P is M.
      <u>All M is S.</u>
      So All S is P.
   c. Invalid
   d. Illicit minor: the minor term, as predicate of an affirmative proposition is particular in the premiss, but universal in the conclusion.
2. a. All who think a lot push the hair out of their heads.
      <u>All philosophers think a lot.</u>

So All philosophers have beards.
b.　All M is Q.
　　All S is M.
　　So All S is P.
c.　Invalid
d.　Four terms; and the predicate of the conclusion does not appear in the premisses.

3.　a.　All who wear black sweaters are serious.
　　　All existentialists wear black sweaters.
　　　So No existentialist wears college colors.
　　b.　All M is Q.
　　　All S is M.
　　　So No S is P.
　　c.　Invalid
　　d.　Four terms; the predicate of the conclusion does not appear in the premisses; a negative conclusion cannot follow two affirmative premisses.

4.　a.　Some theologians are rigorous thinkers.
　　　Some theologians are pantheists.
　　　Some pantheists are rigorous thinkers.
　　b.　Some M is P.
　　　Some M is S.
　　　So Some S are P.
　　c.　Invalid.
　　d.　Undistributed middle.

5.　a.　All Platonists are idealists.
　　　No Aristotelian is a Platonist.
　　　So No Aristotelian is an idealist.
　　b.　All M is P.
　　　No S is M.
　　　So No S is P.
　　c.　Invalid.
　　d.　Illicit major: the P-term is particular in the premiss, since it is the predicate of an affirmative proposition, but it is universal in the conclusion, since it is the predicate of a negative premiss.

6.　a.　All Kantians are devoted to an ethic of duty.
　　　All devoted to an ethic of duty are Prussians.
　　　So All Prussians are Kantians.
　　b.　All P are M.

<u>All M are S.</u>
So All S are P.
c. Invalid
d. Illicit minor. The S-term is used as the subject of a universal proposition in the conclusion, but is given as predicate of an affirmative in the premiss and hence is undistributed.

7. a. No empiricist is a Hegelian.
<u>Some Englishman is a Hegelian.</u>
So Some Englishman is not an empiricist.
   b. No P is M.
<u>Some S is M.</u>
So Some S is not P.
   c. Valid
   d. Second figure

8. a. All sceptics are relativists.
<u>All pragmatists are relativists.</u>
So Some pragmatists are sceptics.
   b. All P is M.
<u>All S is M.</u>
So Some S are P.
   c. Invalid
   d. Undistributed middle. The middle is predicate of an affirmative proposition in both premisses.

9. a. All Stalinists are Leninists.
<u>All Leninists are Marxists.</u>
So All Marxists are Stalinists.
   b. All P are M.
<u>All M are S.</u>
So All S are P.
   c. Invalid
   d. Illicit minor. S is given undistributed as predicate of an affirmative proposition and taken as subject of a universal conclusion.

10. a. All that dance on the head of a pin are viruses.
<u>No angel dances on the head of a pin.</u>
So no angel is a virus.
    b. All M is P.
<u>No S is M.</u>
So No S is P.

c. Invalid
d. Illicit major; "virus" is given undistributed as the predicate of an affirmative premiss and taken as distributed as the predicate of a negative conclusion.

2.321 Enthymeme (p. 110)

Answer to *Exercises*

1. a. All animals that romp in the woods have fleas.
   [   ]
   So Some dogs have fleas.
   b. Some dogs romp in the woods.
   c. All M is P.
   Some S is M.
   So Some S is P.
   d. First figure
2. a. [   ]
   No dog is disloyal.
   So No dog abandons his master in the snow.
   b. Whoever abandons his master in the snow is disloyal.
   c. All P is M.
   No S is M.
   So No S is P.
   d. Second figure. Note that to reverse the missing major, saying "All who are disloyal abandon their masters in the snow", would lead to an illicit major.
3. a. All animals that carry brandy are friendly.
   [   ]
   So All St. Bernards carry brandy.
   b. Incompletable.
   c. All P is M.
   [   ]
   So All S is P.
   d. We need an affirmative minor since we have an affirmative conclusion. The predicate of any affirmative is particular. Yet we need both the S-term and the M-term to be distributed, which is impossible.

4. a. [   ]
        <u>All dachshunds hunt in burrows</u>.
        So No dachshund has long legs.
    b. No thing that hunts in burrows has long legs.
    c. No M is P.
        <u>All S is M,</u>
        So No S is P.
    d. First figure (or, if the converse of the major is used, the
        second figure)
5. a. [   ]
        <u>All animals in this clinic are dogs,</u>
        So Some dogs are near-sighted.
    b. All animals in this clinic are near-sighted. Note that "All
        near-sighted creatures are animals in this clinic" would
        also work formally, but is preposterous.
    c. All M is P.
        <u>All M is S,</u>
        So Some S is P.
    d. Third figure.

2.33 The Hypothetical Syllogism (p. 112)

Answers to *Exercises*

1. Invalid. "Not both" admits also neither, so the premisses
    are inconclusive.
2. Invalid. The premisses, according to *modus tollendo
    tollens,* give "This is not lead."
3. Valid. *Modus ponendo tollens.*
4. Invalid. A non-exclusive "or" allows not just one but both
    alternatives to be true.
5. Valid. *Modus ponendo ponens.*
6. Valid. *Modus tollendo ponens.*
7. Valid. *Modus tollendo tollens.*
8. Invalid. A conditional statement is compatible with the
    consequent being true under another condition, so the
    premisses are inconclusive.

2.35 Truth Tables (p. 115)

Answers to *Exercises*

1.

Blakeley walks.	Colbert singles.	Satty homers.
T	T	T
F	T	T
T	F	T
F	F	T
T	T	F
F	T	F
T	F	F
F	F	F

Blakeley walks.	Colbert singles.	Blakeley walks and Colbert singles.
T	T	T
F	T	F
T	F	F
F	F	F

(Blakeley walks and Colbert singles) implies	Satty homers.	
T	T	T
F	T	T
F	T	T
F	T	T
T	F*	F
F	T	F
F	T	F
F	T	F

The complex is false when Blakeley does walk, Colbert does single, but Satty doesn't get a run.

2.

Satty homers.	Colbert (does) single.	C. doesn't single or S. homers.	
T	T	F	T
F	T	F	F
T	F	T	T
F	F	T	T

Complex is false when Satty doesn't homer, and Colbert singles.

3.

Blakeley walks.	Blakeley doesn't walk.	Blakeley walks or he doesn't.
T	F	T
F	T	T

The statement is always true; it is a tautology.

194

4. Colbert singles. C. fails to single.   C. singles and fails to single.

T	F	F
F	T	F

The statement involves a contradiction and is always false.

5.Satty homers. Blakeley walks. Colbert singles. B. walks or C. singles.

T	T	T	T
F	T	T	T
T	F	T	T
F	F	T	T
T	T	F	T
F	T	F	T
T	F	F	F
F	F	F	F

Satty homers if and only if (Blakeley walks or Colbert singles).

T	T	T
F	F	T
T	T	T
F	F	T
T	T	T
F	F	T
T	F	F
F	T	F

The complex proposition is consistent, neither valid nor inconsistent.
Blakeley walks. Colbert singles. Satty homers. B. walks and C singles.

T	T	T	T
F	T	T	F
T	F	T	F
F	F	T	F
T	T	F	T
F	T	F	F
T	F	F	F
F	F	F	F

(Blakeley walks and Colbert singles) implies Satty homers.

T	T	T
F	T	T
F	T	T
F	T	T
T	F	F
F	T	F
F	T	F
F	T	F

The complex is false only when Blakeley does walk, Colbert does single, but
Satty fails to hit a homer.

2.4   Induction (p. 139)

Answers to *Exercises*

1. We could use the method of difference. First, to see if the average LSAT, GRE, or GMAT scores of students who had taken logic were higher than those of otherwise similar persons who had not. One could refine this by using the method of joint variation to show whether more logic courses relate to better LSAT, GRE, or GMAT scores. Or else one could try to correlate higher logic grades to better LSAT, GRE, or GMAT scores. If one could isolate a group that had taken LSAT, GRE, or GMAT before and again after the logic course, this version of the method of differences would be best of all.

2. No. The American Civil War is a single though complex event. Induction is a process of generalization. Some have tried to make generalizations about civil wars, about revolutions, etc., but even this is questionable, given the uniqueness of each human situation.

3. Using the canon of differences one might get identical twins, preferably ones who do the same type of work, one of whom consumes a great deal of salt, while the other does not. In practice we work backwards. Among heart patients we look for twins, among the siblings, we seek some free of heart disease, then we inquire about their diet. Since almost everybody uses salt, we would not be able to use the method of agreement. Since salt can be used in greater or lesser quantities, we might also try the method of joint variation.

4. One might study persons free of ulcers and look for absence of relevant factors to identify sufficient causes of ulcers, and then look at persons with ulcers to detect common elements. It is unlikely that we would be able to work up to the joint method of differences and agreement because there does not seem to be one necessary and sufficient cause of ulcers.

Appendix 4

Programs

Syllogistics can be automated in a number of ways. This first version is a BASIC program that runs on the VAX (versions 3.5 and following of VMS). It is crudely constructed to make the program flow evident. The reader should feel free to revise it with less GOTOs and more subroutines.

```
50    REM  REVISION 2.0/ COPYRIGHT BY T.J. BLAKELEY
60    PRINT "DO YOU WANT TO SKIP THE INSTRUCTIONS?"
70    PRINT "(PLEASE TYPE 'YES' OR 'NO') "
80    INPUT F$
90    IF F$<>'YES' AND F$<>'NO' THEN GOTO 70
100   IF F$='YES' THEN GOTO 2650
110   PRINT "ACCORDING TO ARISTOTLE'S 'PRIOR ANALYTICS',
      THERE ARE"
120   PRINT TAB(10) ;"THREE BASIC SYLLOGISTIC FIGURES."
130   PRINT "THIS PROGRAM IS DESIGNED TO TELL YOU THE
      FIGURE AND MODE OF"
140   PRINT TAB(5) ;"THE VALID SYLLOGISMS",
160   PRINT TAB(28) ;"NOTE!!!"
170   PRINT "THE  ?  THAT APPEARS IN THE INSTRUCTIONS
      ALLOWS YOU TO READ "
180   PRINT TAB(5) ;"AT YOUR OWN PACE.  WHEN YOU ARE
      READY TO GO ON,"
190   PRINT TAB(10) ;"SIMPLY TOUCH THE 'RETURN' KEY."
200   INPUT G
210   PRINT TAB(5) ;"YOUR REASONING MUST CONSIST OF
      THREE SIMPLE SENTENCES--"
220   PRINT TAB(3) ;"NEITHER COMPLEX NOR COMPOUND."
230   PRINT TAB(10) ;"THERE ARE TWO PREMISSES AND A
```

CONCLUSION."
240   PRINT "IF YOU WOULD LIKE ASSISTANCE IN REDUCING SENTENCES TO SIMPLIFIED FORM,"
250   PRINT "PLEASE TYPE 'HELP' AND TOUCH THE 'RETURN' KEY."
260   PRINT TAB(15) ;"(OTHERWISE, PLEASE JUST TOUCH 'RETURN')"
270   INPUT R$
280   IF R$='HELP' THEN GOSUB 18800
290   PRINT\PRINT
300   PRINT "EACH SENTENCE MUST BE COMPOSED OF FOUR COMPONENTS,"
310   PRINT TAB(20) ;"IN EXACT ORDER,"
320   PRINT TAB(22) ;"AS FOLLOWS:"
330   PRINT TAB(10) ;"1. A QUANTIFIER: 'ALL' OR 'SOME' OR 'NO' (ONLY!)"
340   PRINT TAB(10) ;"2. A SUBJECT, WHICH MUST BE A SINGLE TERM"
350   PRINT TAB(10) ;"3. A SIGN OF EQUALITY (=) OR OF INEQUALITY (<>)"
360   PRINT TAB(10) ;"4. A PREDICATE, WHICH MUST BE A SINGLE TERM"
380   PRINT "(NOTE: A 'SINGLE TERM' IS NOT NECESSARILY A 'SINGLE WORD':"
390   PRINT TAB(3) ;"'RATIONAL ANIMAL',FOR EXAMPLE, IS A SINGLE TERM)"
400   INPUT X
410   PRINT "BOTH OCCURRENCES OF SUBJECT-TERM MUST BE EXACTLY THE SAME; AND"
420   PRINT TAB(5) ;"BOTH OBJECT-TERMS MUST BE RIGIDLY IDENTICAL."
430   PRINT\PRINT\PRINT
440   PRINT TAB(25) ;"FOR EXAMPLE:"
450   PRINT
460   PRINT TAB(20) ;"ALL"; TAB(40) ;"(PUSH 'RETURN')"
470   PRINT TAB(20) ;"HUMANS"; TAB(40) ;"(PUSH 'RETURN')"
480   PRINT TAB(20) ;"="; TAB(40) ;"(PUSH 'RETURN')"
500   PRINT TAB(20) ;"RATIONAL"; TAB(40) ;"(PUSH 'RETURN')"
505   PRINT

```
510   PRINT TAB(20) ;"ALL"; TAB(40) ;"(PUSH 'RETURN')"
520   PRINT TAB(20) ;"PROGRAMMERS"; TAB(40) ;"(PUSH
      'RETURN')"
530   PRINT TAB(20) ;"="; TAB(40) ;"(PUSH 'RETURN')"
540   PRINT TAB(20) ;"HUMANS"; TAB(40) ;"(PUSH 'RETURN')"
550   PRINT
560   PRINT TAB(20) ;"ALL"; TAB(40) ;"(PUSH 'RETURN')"
570   PRINT TAB(20) ;"PROGRAMMERS"; TAB(40) ;"(PUSH
      'RETURN')"
580   PRINT TAB(20) ;"="; TAB(40) ;"(PUSH 'RETURN')"
590   PRINT TAB(20) ;"RATIONAL"; TAB(40) ;"(PUSH 'RETURN')"
600   INPUT G
610   PRINT "WHEN YOU SEE THE FIRST PROMPT, PLEASE TYPE
      THE QUANTIFIER OF YOUR"
620   PRINT TAB(3) ;"FIRST PREMISS, THEN HIT 'RETURN'."
630   PRINT TAB(20) ;"FOR INSTANCE,"
640   PRINT "USING THE ABOVE EXAMPLE, WE WOULD TYPE
      'ALL' AND HIT 'RETURN'."
650   INPUT R
660   PRINT "WHEN THE SECOND PROMPT APPEARS, TYPE THE
      SUBJECT OF YOUR FIRST"
670   PRINT "PROPOSITION AND HIT 'RETURN'."
680   PRINT TAB(25) ;"YOU WILL SEE A TOTAL OF TWELVE
      PROMPTS:"
690   PRINT TAB(28) ;"FOUR TERMS IN EACH OF THE THREE
      PREMISSES."
700   PRINT "(NOTE: PLEASE RENDER 'IS' AS '=' AND 'IS NOT' AS
      '<>'; AND"
710   PRINT TAB(5) ;"NO DISTINCTION SHOULD BE MADE
      BETWEEN SINGULAR AND PLURAL)"
720   INPUT T
2650  PRINT TAB(5) ;"NOW, PLEASE BEGIN TYPING IN THE
      TWELVE TERMS "
2651  REM HOW VARIABLES CORRELATE WITH TERMS
2652  REM                          A$ B$ C$ D$
2654  REM                          E$ F$ G$ H$
2656  REM                          I$  J$  K$ L$
2680  PRINT
2700  INPUT A$,B$,C$,D$,E$,F$,G$,H$,I$,J$,K$,L$
```

2702 REM *FIRST, ELIMINATE MISQUANTIFICATIONS,*
      MISQUALIFICATIONS, MISCOPULATIONS AND BLANK
      TERM-SLOTS
2705 IF A$<>'ALL' AND A$<>'SOME' AND A$<>'NO' THEN GOTO
      19000
2726 IF C$<>'=' AND C$<>'<>' THEN GOTO 19000
2745 IF E$<>'ALL' AND E$<>'SOME' AND E$<>'NO' THEN GOTO
      19000
2766 IF G$<>'=' AND G$<>'<>' THEN GOTO 19000
2785 IF I$<>'ALL' AND I$<>'SOME' AND I$<>'NO' THEN GOTO
      19000
2803 IF K$<>'=' AND K$<>'<>' THEN GOTO 19000
2804 IF A$=' 'OR B$=' 'OR C$=' 'OR D$=' 'OR E$=' 'OR F$=' 'OR G$='
      'OR H$=' 'OR I$=' 'OR J$=' 'OR K$=' 'OR L$=' ' THEN GOTO
      19000
11722 REM *CHECK FOR DOUBLE NEGATION AND DOUBLE*
      PARTICULARIZATION
11725 IF C$='<>' AND G$='<>' THEN 18000
11730 IF C$='<>' AND E$='NO' THEN 18000
11735 IF A$='NO' AND E$='NO' THEN 18000
11740 IF A$='NO' AND G$='<>' THEN 18000
11745 IF A$='SOME' AND E$='SOME' THEN 18200
11750 IF A$='ALL'AND E$='ALL'AND D$=H$ THEN GOTO 18600
11760 IF A$='ALL'AND E$='SOME'AND D$=H$ AND G$='=' THEN
      GOTO 18600
11770 IF A$='ALL'AND E$='SOME'AND F$=D$ THEN GOTO 18600
11780 IF A$='SOME'AND E$='ALL'AND B$=H$ THEN GOTO 18600
11790 IF A$='SOME'AND E$='ALL'AND D$=H$ AND C$='=' THEN
      GOTO 18600
11792 IF A$='ALL'AND E$='SOME'AND I$='ALL' THEN GOTO 18700
11794 IF A$='SOME'AND E$='ALL'AND I$='ALL' THEN GOTO 18700
11796 IF A$='NO'AND I$='ALL' THEN GOTO 18700
11797 IF E$='NO' AND I$='ALL' THEN GOTO 18700
11798 IF C$ = '<>' AND K$='=' THEN GOTO 18700
11799 IF G$='<>' AND K$='=' THEN GOTO 18700
11800 IF A$='NO' AND K$='=' AND I$='SOME' THEN GOTO 18700
11801 IF A$='SOME' AND I$<>'SOME' THEN GOTO 18700
11803 IF E$='SOME' AND I$<>'SOME' THEN GOTO 18700
11805 IF E$='NO' AND K$='=' AND I$='SOME' THEN GOTO 18700

```
11810 REM BEGIN HERE THE VALID SYLLOGISTIC FORMS
11820 IF A$='ALL'AND E$='ALL'AND I$='ALL'AND B$=H$ AND F$=J$
      AND D$=L$ AND C$='='AND G$='='AND K$='=' THEN GOTO
      15000
11830 IF A$='ALL'AND E$='ALL'AND I$='SOME'AND B$=H$ AND
      F$=J$ AND D$=L$ AND C$='='AND G$='='AND K$='=' THEN
      GOTO 17705
11840 IF A$='ALL'AND E$='SOME'AND I$='SOME'AND B$=H$ AND
      F$=J$ AND D$=L$ AND C$='='AND G$='='AND K$='=' THEN
      GOTO 15600
11850 IF A$='NO'AND E$='ALL'AND I$='NO'AND B$=H$ AND F$=J$
      AND D$=L$ AND C$='='AND G$='='AND K$='=' THEN GOTO
      16200
11860 IF A$='NO'AND E$='ALL'AND I$='SOME'AND B$=H$ AND
      F$=J$ AND D$=L$ AND C$='='AND G$='='AND K$='<>' THEN
      GOTO 17715
11870 IF A$='NO'AND E$='SOME'AND I$='SOME'AND B$=H$ AND
      F$=J$ AND D$=L$ AND C$='='AND G$='='AND K$='<>' THEN
      GOTO 17000
11880 IF A$='ALL'AND E$='NO'AND I$='NO'AND D$=H$ AND F$=J$
      AND B$=L$ AND C$='='AND G$='='AND K$='=' THEN GOTO
      15400
11890 IF A$='ALL'AND E$='NO'AND I$='SOME'AND D$=H$ AND
      F$=J$ AND B$=L$ AND C$='='AND G$='='AND K$='<>' THEN
      GOTO 17725
11900 IF A$='ALL'AND E$='SOME'AND I$='SOME'AND D$=H$ AND
      F$=J$ AND B$=L$ AND C$='='AND G$='<>'AND K$='<>' THEN
      GOTO 15800
11910 IF A$='NO'AND E$='ALL'AND I$='NO'AND D$=H$ AND F$=J$
      AND B$=L$ AND C$='='AND G$='='AND K$='=' THEN GOTO
      16400
11920 IF A$='NO'AND E$='ALL'AND I$='SOME'AND D$=H$ AND
      F$=J$ AND B$=L$ AND C$='='AND G$='='AND K$='<>' THEN
      GOTO 17735
11930 IF A$='NO'AND E$='SOME'AND I$='SOME'AND D$=H$ AND
      F$=J$ AND B$=L$ AND C$='='AND G$='='AND K$='<>' THEN
      GOTO 17200
12000 IF A$='ALL'AND E$='ALL'AND I$='SOME'AND B$=F$ AND
      H$=J$ AND D$=L$ AND C$='='AND G$='='AND K$='=' THEN
```

```
        GOTO 15200
12100 IF A$='ALL'AND E$='SOME'AND I$='SOME'AND B$=F$ AND
      H$=J$ AND D$=L$ AND C$='='AND G$='='AND K$='=' THEN
      GOTO 16000
12200 IF A$='NO'AND E$='ALL'AND I$='SOME'AND B$=F$ AND
      H$=J$ AND D$=L$ AND C$='='AND G$='='AND K$='<>' THEN
      GOTO 17745
12300 IF A$='NO'AND E$='SOME'AND I$='SOME'AND B$=F$ AND
      H$=J$ AND D$=L$ AND C$='='AND G$='='AND K$='<>' THEN
      GOTO 17400
12400 IF A$='SOME'AND E$='ALL'AND I$='SOME'AND B$=F$ AND
      H$=J$ AND D$=L$ AND C$='='AND G$='='AND K$='=' THEN
      GOTO 16800
12500 IF A$='SOME'AND E$='ALL'AND I$='SOME'AND B$=F$ AND
      H$=J$ AND D$=L$ AND C$='<>'AND G$='='AND K$='<>' THEN
      GOTO 16600
12600 IF A$='ALL'AND E$='ALL'AND I$='SOME'AND F$=D$ AND
      B$=L$ AND H$=J$ AND C$='='AND G$='='AND K$='=' THEN
      GOTO 18949
12700 IF A$='ALL'AND E$='NO'AND I$='NO'AND F$=D$ AND B$=L$
      AND H$=J$ AND C$='='AND G$='='AND K$='=' THEN GOTO
      18949
12800 IF A$='ALL'AND E$='NO'AND I$='SOME'AND F$=D$ AND
      B$=L$ AND H$=J$ AND C$='='AND G$='=' AND K$='<>' THEN
      GOTO 18949
12850 IF A$='NO' AND E$='ALL'AND I$='NO'AND F$=D$ AND B$=L$
      AND H$=J$ AND C$='='AND G$='='AND K$='=' THEN GOTO
      18949
12900 IF A$='NO'AND E$='ALL'AND I$='SOME'AND F$=D$ AND
      B$=L$ AND H$=J$ AND C$='='AND G$='='AND K$='<>' THEN
      GOTO 18949
12950 IF A$='NO'AND E$='SOME'AND I$='SOME'AND F$=D$ AND
      B$=L$ AND H$=J$ AND C$='='AND G$='='AND K$='<>' THEN
      GOTO 18949
13000 IF A$='SOME'AND E$='ALL'AND I$='SOME'AND F$=D$ AND
      B$=L$ AND H$=J$ AND C$='='AND G$='='AND K$='=' THEN
      GOTO 18949
13050 IF A$='ALL'AND E$='ALL'AND I$='SOME'AND B$=H$ AND
      F$=L$ AND D$=J$ AND C$='='AND G$='='AND K$='=' THEN
```

GOTO 18949

13100 IF A$='ALL'AND E$='NO'AND I$='SOME'AND B$=H$ AND
F$=L$ AND D$=J$ AND C$='='AND G$='='AND K$='<>' THEN
GOTO 18949

13150 IF A$='ALL'AND E$='SOME'AND I$='SOME'AND B$=H$ AND
F$=L$ AND D$=J$ AND C$='='AND G$='='AND K$='=' THEN
GOTO 18949

13200 IF A$='NO'AND E$='ALL'AND I$='NO'AND B$=H$ AND F$=L$
AND D$=J$ AND C$='='AND G$='='AND K$='=' THEN GOTO
18949

13250 IF A$='NO'AND E$='ALL'AND I$='SOME'AND B$=H$ AND
F$=L$ AND D$=J$ AND C$='='AND G$='='AND K$='<>' THEN
GOTO 18949

13300 IF A$='SOME'AND E$='NO'AND I$='SOME'AND B$=H$ AND
F$=L$ AND D$=J$ AND C$='='AND G$='='AND K$='<>' THEN
GOTO 18949

13810 IF A$='ALL'AND E$='ALL'AND I$='NO'AND B$=H$ AND F$=J$
AND D$=L$ AND C$='='AND G$='='AND K$='=' THEN GOTO
18400

13820 IF A$='ALL'AND E$='ALL'AND I$='SOME'AND B$=H$ AND
F$=J$ AND D$=L$ AND C$='='AND G$='='AND K$='<>' THEN
GOTO 18400

13830 IF A$='ALL'AND E$='NO'AND I$='NO'AND B$=H$ AND F$=J$
AND D$=L$ AND C$='='AND G$='='AND K$='=' THEN GOTO
18400

13840 IF A$='ALL'AND E$='SOME' AND I$='SOME'AND B$=H$ AND
F$=J$ AND D$=L$ AND C$='='AND G$='='AND K$='<>'THEN
GOTO 18400

13850 IF A$='ALL'AND E$='SOME'AND I$='SOME'AND B$=H$ AND
F$=J$ AND D$=L$ AND C$='='AND G$='<>'AND K$='<>' THEN
GOTO 18400

13860 IF A$='SOME'AND E$='NO'AND I$='SOME'AND B$=H$ AND
F$=J$ AND D$=L$ AND C$='='AND G$='='AND K$='<>'THEN
GOTO 18400

13870 IF A$='ALL'AND E$='ALL'AND I$='ALL'AND B$=F$ AND H$=J$
AND D$=L$ AND C$='='AND G$='='AND K$='=' THEN GOTO
18400

13880 IF A$='ALL'AND E$='ALL'AND I$='NO'AND B$=F$ AND H$=J$
AND D$=L$ AND C$='='AND G$='='AND K$='=' THEN GOTO

18400

13890 IF A$='ALL' AND E$='ALL'AND I$='SOME'AND B$=F$ AND
H$=J$ AND D$=L$ AND C$='='AND G$='='AND K$='<>' THEN
GOTO 18400

13900 IF A$='ALL'AND E$='NO'AND I$='NO'AND B$=F$ AND H$=J$
AND D$=L$ AND C$='='AND G$='='AND K$='=' THEN GOTO
18400

13910 IF A$='ALL'AND E$='NO'AND I$='SOME'AND B$=F$ AND
H$=J$ AND D$=L$ AND C$='='AND G$='='AND K$='<>' THEN
GOTO 18400

13915 IF A$='ALL'AND E$='NO'AND I$='SOME'AND B$=H$ AND
F$=J$ AND D$=L$ AND C$='='AND G$='='AND K$='<>' THEN
GOTO 18400

13920 IF A$='ALL'AND E$='SOME'AND I$='SOME'AND B$=F$ AND
H$=J$ AND D$=L$ AND C$='='AND G$='='AND K$='<>' THEN
GOTO 18400

13930 IF A$='ALL'AND E$='SOME'AND I$='SOME'AND B$=F$ AND
H$=J$ AND D$=L$ AND C$='='AND G$='<>'AND K$='<>' THEN
GOTO 18400

13940 IF A$='NO'AND E$='ALL'AND I$='NO'AND B$=F$ AND H$=J$
AND D$=L$ AND C$='='AND G$='='AND K$='=' THEN GOTO
18400

13950 IF A$='SOME'AND E$='ALL'AND I$='SOME'AND B$=F$ AND
H$=J$ AND D$=L$ AND C$='='AND G$='='AND K$='<>' THEN
GOTO 18400

13960 IF A$='SOME'AND E$='NO'AND I$='SOME'AND B$=F$ AND
H$=J$ AND D$=L$ AND C$='='AND G$='='AND K$='<>' THEN
GOTO 18400

13970 IF A$='SOME'AND E$='NO'AND I$='SOME'AND D$=H$ AND
F$=J$ AND B$=L$ AND C$='='AND G$='='AND K$='<>' THEN
GOTO 18400

13980 IF A$='SOME'AND E$='ALL'AND I$='SOME'AND D$=H$ AND
F$=J$ AND B$=L$ AND C$='<>'AND G$='='AND K$='<>' THEN
GOTO 18400

13990 IF A$='ALL'AND E$='ALL'AND I$='ALL'AND F$=D$ AND
B$=L$ AND H$=J$ AND C$='='AND G$='='AND K$='=' THEN
GOTO 18400

14000 IF A$='ALL'AND E$='ALL'AND I$='NO'AND F$=D$ AND B$=L$
AND H$=J$ AND C$='='AND G$='='AND K$='=' THEN GOTO

```
           18400
14010 IF A$='SOME'AND E$='ALL' AND I$='SOME'AND F$=D$ AND
      B$=L$ AND H$=J$ AND C$='='AND G$='='AND K$='<>' THEN
      GOTO 18400
14020 IF A$='SOME'AND E$='NO'AND I$='SOME'AND F$=D$ AND
      B$=L$ AND H$=J$ AND C$='='AND G$='='AND K$='<>' THEN
      GOTO 18400
14030 IF A$='SOME'AND E$='ALL'AND I$='SOME'AND F$=D$ AND
      B$=L$ AND H$=J$ AND C$='<>'AND G$='='AND K$='<>' THEN
      GOTO 18400
14040 IF A$='ALL'AND E$='ALL'AND I$='ALL'AND B$=H$ AND
      F$=L$ AND D$=J$ AND C$='='AND G$='='AND K$='=' THEN
      GOTO 18400
14050 IF A$='ALL'AND E$='ALL'AND I$='NO'AND B$=H$ AND F$=L$
      AND D$=J$ AND C$='='AND G$='='AND K$='=' THEN GOTO
      18400
14060 IF A$='ALL'AND E$='ALL'AND I$='SOME'AND B$=H$ AND
      F$=L$ AND D$=J$ AND C$='='AND G$='='AND K$='<>' THEN
      GOTO 18400
14070 IF A$='ALL'AND E$='SOME'AND I$='SOME'AND B$=H$ AND
      F$=L$ AND D$=J$ AND C$='='AND G$='='AND K$='<>' THEN
      GOTO 18400
14080 IF A$='ALL'AND E$='SOME'AND I$='SOME'AND B$=H$ AND
      F$=L$ AND D$=J$ AND C$='='AND G$='<>'AND K$='<>' THEN
      GOTO 18400
14090 IF A$='NO'AND E$='SOME'AND I$='SOME'AND B$=H$ AND
      F$=L$ AND D$=J$ AND C$='='AND G$='='AND K$='<>' THEN
      GOTO 18400
14100 IF A$='ALL'AND E$='ALL'AND I$='ALL'AND B$=F$ AND D$=J$
      AND H$=L$ THEN GOTO 18400
14410 REM DEFAULT FOLLOWS (14420-14430)
14420 PRINT "THIS IS NOT IN ONE OF ARISTOTLE'S THREE
      FIGURES "
14430 GOTO 19000
14490 REM NOW, NAME THE VALID SYLLOGISMS
15000 PRINT
15100 PRINT 'THIS VALID SYLLOGISM IS BARBARA IN FIGURE
      1 (CF.25B39)'
15105 GOSUB 17872
```

```
15150  GOTO 19000
15200  PRINT
15300  PRINT 'THIS VALID SYLLOGISM IS DARAPTI IN FIGURE
       3 (CF.28A18)'
15305  GOSUB 17892
15350  GOTO 19000
15400  PRINT
15500  PRINT 'THIS VALID SYLLOGISM IS CAMESTRES IN
       FIGURE 2 (CF.27A9)'
15505  GOSUB 17882
15550  GOTO 19000
15600  PRINT
15700  PRINT 'THIS VALID SYLLOGISM IS DARII IN FIGURE 1
       (CF.26A23)'
15705  GOSUB 17872
15750  GOTO 19000
15800  PRINT
15900  PRINT 'THIS VALID SYLLOGISM IS BAROCO IN FIGURE 2
       (CF.27A37)'
15905  GOSUB 17882
15950  GOTO 19000
16000  PRINT
16100  PRINT 'THIS VALID SYLLOGISM IS DATISI IN FIGURE 3
       (CF.28B13)'
16105  GOSUB 17892
16150  GOTO 19000
16200  PRINT
16300  PRINT 'THIS VALID SYLLOGISM IS CELARENT IN
       FIGURE 1 (CF.26A1)'
16305  GOSUB 17872
16350  GOTO 19000
16400  PRINT
16500  PRINT 'THIS VALID SYLLOGISM IS CESARE IN FIGURE 2
       (CF.27A5)'
16505  GOSUB 17882
16550  GOTO 19000
16600  PRINT
16700  PRINT 'THIS VALID SYLLOGISM IS BOCARDO IN FIGURE
       3 (CF.28B17)'
```

```
16705 GOSUB 17892
16750 GOTO 19000
16800 PRINT
16900 PRINT 'THIS VALID SYLLOGISM IS DISAMIS IN FIGURE 3
      (CF.28B8)'
16905 GOSUB 17892
16950 GOTO 19000
17000 PRINT
17100 PRINT 'THIS VALID SYLLOGISM IS FERIO IN FIGURE 1
      (CF.26A25)'
17105 GOSUB 17872
17150 GOTO 19000
17200 PRINT
17300 PRINT 'THIS VALID SYLLOGISM IS FESTINO IN FIGURE 2
      (CF.27A32)'
17305 GOSUB 17882
17350 GOTO 19000
17400 PRINT
17600 PRINT 'THIS VALID SYLLOGISM IS FERISON IN FIGURE 3
      (CF.28B33)'
17605 GOSUB 17892
17700 GOTO 19000
17705 PRINT 'THIS VALID SYLLOGISM IS BARBARI IN FIGURE
      1 (=SUBALTERNATE)'
17707 GOSUB 17872
17710 GOTO 19000
17715 PRINT 'THIS VALID SYLLOGISM IS CELARONT IN
      FIGURE 1 (=SUBALTERNATE)'
17717 GOSUB 17872
17720 GOTO 19000
17725 PRINT 'THIS VALID SYLLOGISM IS CAMESTROP IN
      FIGURE 2 (=SUBALTERNATE)'
17727 GOSUB 17882
17730 GOTO 19000
17735 PRINT 'THIS VALID SYLLOGISM IS CESARO IN FIGURE 2
      (=SUBALTERNATE)'
17737 GOSUB 17882
17740 GOTO 19000
17745 PRINT 'THIS VALID SYLLOGISM IS FELAPTON IN FIGURE
```

```
              3 (28A26)'
17747  GOSUB 17892
17750  GOTO 19000
17760  REM FORMAL STRUCTURES
17770  REM The ' ' = ring a bell
17872  PRINT    ! FIGURE 1
17874  PRINT TAB(10) ;'M - P'; TAB(40);' '
17876  PRINT TAB(10) ;'S - M'; TAB(45);' '
17878  PRINT TAB(10) ;'S - P'; TAB(50);' '
17880  RETURN
17882  PRINT    ! FIGURE 2
17884  PRINT TAB(10) ;'P - M'; TAB(50) ;' '
17886  PRINT TAB(10) ;'S - M'; TAB(55) ;' '
17888  PRINT TAB(10) ;'S - P'; TAB(60) ;' '
17890  RETURN
17892  PRINT    ! FIGURE 3
17894  PRINT TAB(10) ;'M - P'; TAB(60) ;' '
17896  PRINT TAB(10) ;'M - S'; TAB(65) ;' '
17898  PRINT TAB(10) ;'S - P'; TAB(70) ;' '
17900  RETURN
17990  REM   BASIC REASONS FOR INVALIDATION (EXCLUDING
       THE DEFAULT!)
18000  PRINT
18100  PRINT 'NO  CONCLUSION  FOLLOWS  FROM  TWO
       NEGATIVE PREMISES BECAUSE'
18102  PRINT 'TWO EXCLUSIONS PROVIDE NO (NECESSARY)
       INCLUSION'
18150  GOTO 19000
18200  PRINT
18300  PRINT 'NO  CONCLUSION  FOLLOWS  FROM  TWO
       PARTICULAR PREMISES SINCE'
18302  PRINT 'TWO PARTIAL INCLUSIONS DO NOT NECESSARILY
       INCLUDE EACH OTHER'
18304  GOTO 19000
18400  PRINT\PRINT
18405  PRINT 'THIS  SYLLOGISM  IS  INVALID  BECAUSE
       SOMETHING DISTRIBUTED IN THE'
18406  PRINT 'CONCLUSION  IS  UNDISTRIBUTED  IN  THE
       PREMISSES.'
```

18408 PRINT TAB(15) ;"IN OTHER WORDS, IF A TERM OCCURS UNIVERSALLY IN THE CONCLUSION"
18410 PRINT "BUT PARTICULARLY IN A PREMISS, THEN THE CONCLUSION WOULD BE SAYING"
18412 PRINT TAB(3) ;"MORE THAN THE PREMISS WARRANTS."
18500 GOTO 19000
18600 PRINT 'THIS SYLLOGISM IS INVALID BECAUSE NEITHER OCCURRENCE OF THE'
18602 PRINT 'MIDDLE TERM IS DISTRIBUTED.'
18604 PRINT TAB(20) ;"IN OTHER WORDS, AT LEAST ONE OCCURRENCE OF THE MIDDLE TERM"
18606 PRINT "MUST BE UNIVERSAL; BECAUSE THE CO-OCCURRENCE OF TWO PARTICULARS"
18608 PRINT TAB(8) ;"CAN GUARANTEE NOTHING. AND, IF THE MIDDLE TERM IS AMBIGUOUS,"
18610 PRINT TAB(5) ;"THE SYLLOGISM HAS FOUR TERMS AND FAILS."
18620 GOTO 19000
18700 PRINT 'THIS SYLLOGISM VIOLATES THE RULE THAT "THE CONCLUSION FOLLOWS'
18702 PRINT 'THE WEAKER PART". IN OTHER WORDS, IF THERE IS A NEGATIVE PREMISS,'
18703 PRINT 'THE CONCLUSION CANNOT BE POSITIVE; IF THERE IS A PARTICULAR'
18704 PRINT 'PREMISS, THE CONCLUSION CANNOT BE UNIVERSAL.'
18798 GOTO 19000
18800 PRINT TAB(20) ;"INDICATIONS ON SIMPLIFYING SENTENCES FOR SYLLOGISTICS,"
18802 PRINT TAB(22) ;"DRAWN FROM THE PROGRAM WFF001."
18804 PRINT
18806 PRINT "SINCE ONLY 'ALL', 'SOME', 'NO' CAN APPEAR AS QUANTIFIERS,"
18808 PRINT TAB(10) ;"'EACH' AND 'EVERY' SHOULD BE RENDERED AS 'ALL'."
18810 PRINT
18812 PRINT TAB(8) ;"'THIS' CANNOT APPEAR; NOR CAN PROPER NAMES--"

18814 PRINT "BECAUSE WHAT IS 'THIS' CAN BELONG TO WHAT IS 'ALL' OR 'SOME'"
18816 PRINT TAB(10) ;"BUT THE INVERSE IS NOT THE CASE."
18818 PRINT "ACCORDING TO 43A25-43B39, SCIENCE IS ABOUT WHAT IS ESSENTIAL,"
18820 PRINT TAB(10) ;"NOT ABOUT WHAT IS INDIVIDUAL AND ACCIDENTAL."
18822 INPUT F
18824 PRINT "FOUR SORTS OF PROPOSITION OCCUR IN TRADITIONAL SYLLOGISTICS,"
18826 PRINT TAB(5) ;"EACH REPRESENTED BY A LETTER WHICH OCCURS AS A VOWEL"
18828 PRINT TAB(8) ;"IN THE MNEMONIC TERMS."
18830 PRINT\PRINT
18832 PRINT TAB(30) ;'"A" = UNIVERSAL AFFIRMATIVE (ALL)'
18834 PRINT TAB(30) ;'"I" = PARTICULAR AFFIRMATIVE (SOME)'
18836 PRINT TAB(30) ;'"E" = UNIVERSAL NEGATIVE (NO)'
18838 PRINT TAB(30) ;'"O" = PARTICULAR NEGATIVE (SOME...IS NOT...)'
18840 PRINT
18842 PRINT 'THUS:'; TAB(20) ;'BARBARA = A + A + A'
18844 PRINT TAB(20) ;'DISAMIS = I + A + I'
18846 INPUT Y
18848 PRINT TAB(10) ;"THERE ARE THREE BASIC SYLLOGISTIC 'FIGURES' (SCHEMATA),"
18850 PRINT "DISTINGUISHED BY THE POSITION OF THE THREE TERMS THAT MAKE THE SYLLOGISM POSSIBLE:"
18851 PRINT
18852 PRINT TAB(10);"FIGURE 1"; TAB(30);"FIGURE 2"; TAB(50) ;"FIGURE 3"
18853 PRINT
18854 PRINT TAB(10) ;"M - P"; TAB(30) ;"P - M"; TAB(50) ;"M - P"
18856 PRINT TAB(10) ;"S - M"; TAB(30) ;"S - M"; TAB(50) ;"M - S"
18858 PRINT TAB(10) ;"S - P"; TAB(30) ;"S - P"; TAB(50) ;"S - P"
18860 PRINT\PRINT
18876 INPUT Q
18878 PRINT TAB(8) ;"THE TWO MOST DIFFICULT HABITS TO ACQUIRE HAVE TO DO WITH:"

18880 PRINT TAB(3) ;"(1) RENDERING COMPLEX AND COMPOUND ENGLISH SENTENCES INTO SUBJECT-PREDICATE FORM, AND"
18882 PRINT TAB(3) ;"(2) WHAT TO DO WITH VARIOUS PHRASES AND CLAUSES--ESPECIALLY RELATIVES."
18884 PRINT\PRINT
18886 PRINT TAB(3) ;"(1) TO RENDER COMPLEX ENGLISH SENTENCES INTO SUBJECT-PREDICATE FORM,"
18888 PRINT "CHANGE THE VERB INTO 'IS' + A PARTICIPLE (WHICH IS A VERBAL ADJECTIVE)"
18890 PRINT
18892 PRINT TAB(3) ;"(2) WHENEVER POSSIBLE, PHRASES AND CLAUSES SHOULD ALSO BE PUT INTO ADJECTIVAL FORM."
18894 INPUT E
18896 PRINT TAB(30) ;"AN EXAMPLE:"
18898 PRINT
18900 PRINT TAB(8) ;"EVERYONE WHO HAS EVER HAD TO REDUCE ORDINARY SENTENCES TO"
18902 PRINT "CANONICAL FORM CAN EMPATHIZE WITH THE POLISH PATRIOTS."
18904 INPUT P
18906 PRINT "STEP 1: PUT 'ALL' FOR 'EVERYONE'"
18908 PRINT
18910 PRINT "STEP 2: PUT 'ABLE TO EMPATHIZE WITH...' FOR 'CAN EMPATHIZE...'."
18912 PRINT
18914 PRINT "STEP 3: PUT 'HAVING TO REDUCE...' FOR 'WHO HAS EVER...'."
18916 INPUT H
18918 PRINT TAB(3) ;"ASSUMING THAT THAT SENTENCE IS THE CONCLUSION OF A SYLLOGISM,"
18920 PRINT "'ALL HAVING TO REDUCE...' IS 'S' AND 'ABLE TO EMPATHIZE...' IS 'P'."
18922 INPUT B
18924 PRINT TAB(10) ;"WE COULD MAKE A 'BARBARA' AS FOLLOWS:"
18926 PRINT
18928 PRINT TAB(1);"ALL THOSE DOING THE VERY DIFFICULT";

```
             TAB(40);"="; TAB(45);"ABLE TO EMPATHIZE..."
18930  PRINT TAB(1);"ALL HAVING TO REDUCE...";   TAB(40);"=";
       TAB(45);"THOSE DOING THE VERY DIFFICULT"
18932  PRINT TAB(1);"ALL HAVING TO REDUCE...";   TAB(40);"=";
       TAB(45);"ABLE TO EMPATHIZE..."
18934  INPUT C
18936  PRINT "IF ALL ELSE FAILS, CALL (617) 471 9052."
18940  PRINT "NOW THAT YOU CAN SIMPLIFY PROPOSITIONS,
       HERE ARE MORE DETAILS ON"
18945  PRINT "PREMISSES AND HOW TO DEAL WITH THEM."
18947  RETURN
18949  PRINT "THIS BELONGS TO FIGURE 4 OR THE INDIRECT
       FIRST AND, THEREFORE,"
18951  PRINT "   NOT TO ARISTOTLE   "
18998  RETURN
19000  PRINT
19100  PRINT "WOULD YOU LIKE TO TRY ANOTHER SYLLOGISM?"
19110  PRINT "(PLEASE TYPE 'YES' OR 'NO')"
19120  INPUT X$
19130  IF X$<>'YES' AND X$<>'NO' THEN GOTO 19110
19300  IF X$='YES' THEN 2650
19999  END
```

*   *   *

Here, in Pascal, is a program that checks for the First Figure
only. The same devices can be used to extend it to the other
figures.

```
program siligism1(input,output);
  VAR majsub,majpred,minsub,minpred,subcon,
    predcon : packed array[1..20]of char;
    quanmaj,quanmin,quancon:packed array[1..5]of char;
      copmaj,copmin,copcon:packed array[1..2]of char;
begin
  writeln;
  writeln('Please make all entries in CAPITAL letters!');
  writeln('Please enter the quantifier of your major premiss: ');
```

```
  readln(quanmaj);
  writeln('...and its subject...');
  readln(majsub);
  writeln('...and its copula...');
  readln(copmaj);
  writeln('...and its predicate: ');
  readln(majpred);
  writeln('Please enter the quantifier of your minor premiss: ');
  readln(quanmin);
  writeln('...and its subject...');
  readln(minsub);
  writeln('...and its copula...');
  readln(copmin);
  writeln('...and its predicate...');
  readln(minpred);
  writeln('Please enter the quantifier of your conclusion: ');
  readln(quancon);
  writeln('...and its subject...');
  readln(subcon);
  writeln('...and its copula...');
  readln(copcon);
  writeln('...and its predicate: ');
  readln(predcon);
if(majsub=minpred)and(minsub=subcon)and(majpred=predcon)
 and(quanmaj='ALL ')and(quanmin='ALL ')and(quancon='ALL ')
 and(copmaj='= ')and(copmin='= ')and(copcon='= ') then
 writeln('This valid',chr(7),' syllogism is BARBARA in Figure 1.') else
if(majsub=minpred)and(minsub=subcon)and(majpred=predcon)
 and(quanmaj='ALL ')and(quanmin='ALL ')and(quancon='SOME ')
 and(copmaj='= ')and(copmin='= ')and(copcon='= ') then
 writeln('This valid syllogism is BARBARI in Figure 1.',chr(7)) else
if(majsub=minpred)and(minsub=subcon)and(majpred=predcon)
 and(quanmaj='ALL ')and(quanmin='SOME ')and(quancon='SOME ')
 and(copmaj='= ')and(copmin='= ')and(copcon='= ') then
 writeln('This valid syllogism is DATISI',chr(7),' in Figure 1.') else
if(majsub=minpred)and(minsub=subcon)and(majpred=predcon)
 and(quanmaj='NO ')and(quanmin='ALL ')and(quancon='NO ')
 and(copmaj='= ')and(copmin='= ')and(copcon='= ') then
  writeln('This valid syllogism ',chr(7),'is CELARENT in Figure 1.')
```

```
else
    if(majsub=minpred)and(minsub=subcon)and(majpred=predcon)
    and(quanmaj='NO  ')and(quanmin='ALL ')and(quancon='SOME ')
    and(copmaj='= ')and(copmin='= ')and(copcon='<>') then
        writeln('This valid syllogism is ',chr(7),'CELARONT in Figure 1.')
else
    if(majsub=minpred)and(minsub=subcon)and(majpred=predcon)
    and(quanmaj='NO  ')and(quanmin='SOME ')and(quancon='SOME ')
    and(copmaj='= ')and(copmin='= ')and(copcon='<>') then
    writeln('This valid syllogism ',chr(7),'is FERIO in Figure 1.') else
    writeln('This is not in Figure 1.')
    end.
```

SELECTED BIBLIOGRAPHY

1) William and Martha Kneale, *The Development of Logic*, Clarendon Press, Oxford, 1962. Most extensive history, primarily of formal logic.

2) I.M.Bochenski, *A History of Formal Logic*, University of Notre Dame, Notre Dame, Indiana. An anthology of texts with ample introduction and commentaries.

3) J. M. Bochenski, *A Preçis of Mathematical Logic*, trans. Otto Bird, Reidel, Dordrecht, Holland, 1959. *The Methods of Contemporary Thought*, trans. Peter Caws, Dordrecht, Holland, 1965. "Mathematical logic" is a 1950's equivalent of "symbolic logic". Both books are complete and succinct in their respective fields of contemporary formal logic and applied logic or methodology.

4) Plato (428/7-348/7), *Phaedrus*. Contains basic statement on definition and classification, topics to which Plato returns continually.

5) Aristotle, *The Organon*, means tool (Aristotle does not use the term "logic"); consists of six works: *Categories, On Interpretation*, (Peri Hermeneias), *Prior Analytics, Posterior Analytics, Topics, On Sophistical Refutations*. The *Prior Analytics* describe the syllogism, which Aristotle regards as the basic element of reasoning, and the *Posterior Analytics* give his theory of science.

6) John of St. Thomas (1589-1644), *The Material Logic of John of St. Thomas: Basic Treatises*, trans. Yves R. Simon, John J. Glanville, G. Donald Hollenhorst, Preface Jacques Maritain, U. of Chicago Press, Chicago, 1955. Translation from *Ars Logica*, part of *Cursus Philosophicus* of the greatest

scholastic systematizer. Also *Outlines of Formal Logic*, trans. Francis C. Wade, S.J., Marquette University, Milwaukee, 1955.

7) John Stuart Mill (1806-1873), *System of Logic, Ratiocinative and Inductive*, 2 vols., University of Toronto, 1974. Published in 1843 at the empiricist, psychologist nadir of formal logic and philosophy of logic, Mill presented what remains the standard textbook treatment of induction.

6) George Boole (1815-1864), *An Investigation of the Laws of Thought*, Dover, New York, corrected reprinting of 1854 edition. Seminal work in renaissance of formal logic and its metamorphosis into symbolic logc.

9) Lewis Carroll (1832-1898), *Symbolic Logic, Game of Logic*, Dover, N.Y., 1958 reprint of separate versions of 1897 and 1887. Applications, problems, and games prepared by Charles Ludwige Dodgson, whose formal and semantic playfulness will live forever in *Alice in Wonderland* and *The Hunting of the Snark*.

10) H.W.B. Joseph (1867-1943) *An Introduction to Logic*, Clarendon Press, Oxford, 2nd. ed. 1916; 6th reprinting 1961. Last major work of traditional logic unaffected by symbolic logic.

11) Alfred North Whitehead (1861-1947) and Bertrand Russell (1872-1970), *Principia Mathematica*, Cambridge U. Press, vol. I, 1st ed. 1910, 2nd, 1925; vol. II, 1st ed. 1912, 2nd ed. 1927; vol. III, 1st ed. 1913, 2nd ed. 1927; reprinted 1957. Mammoth attempt to achieve in symbolic guise Aristotle's ideal of an axiomatic science, specifically by deducing all of formal logic and thence all of arithmetic from five logical postulates. Since the work of Kurt Gödel (b.1906), the attempt to derive arithmetic from logic is usually regarded as impossible.

12) Rudolf Carnap (1891-1970), *The Logical Syntax of Language*, trans. Amethe Smeaton (Countess von Zeppelin) Littlefield, Adams, Paterson, N.J., 1959. *Introduction to Semantics* and *Formalization of Logic*, Harvard University,

Cambridge, Mass., 1942 and 1943, republished together in 1961. Carnap's attempt to formalize semantics can be regarded as a precursor of our work on the syntactic parser.

13) Further reading on BASIC should include *Hands-on Basic* by Herbert D. Peckham (McGraw-Hill) and *The Basic Handbook* by David Lien (Compusoft Publ.).

14) Automatic parsers are benefitting from work being done in PROLOG in Edinburgh and Marseille, but also in Poland, Hungary and Japan. While the classic text is *Programming in PROLOG* by W.F. Clocksin and C.S. Mellish (Springer Verlag), the most relevant to our objectives is *Logic for Natural Language Analysis* by Fernando Pereira (SRI International, Menlo Park, CA), which has many examples and an extensive bibliography.

INDEX

Berkeley

LIBRARY OF THE UNIVERSITY OF CALIFORNIA